great

Vcm

Wcpo.

Wcpo = f (Vcm)

Advances in Industrial Control

Other titles published in this Series:

Radiotherapy Treatment Planning
Olivier Haas

Performance Assessment of Control Loops
Biao Huang and Sirish L. Shah

Data-driven Techniques for Fault Detection and Diagnosis in Chemical Processes
Evan L. Russell, Leo H. Chiang and Richard D. Braatz

Non-linear Model-based Process Control
Rashid M. Ansari and Moses O. Tadé

Nonlinear Identification and Control
Guoping Liu

Digital Controller Implementation and Fragility
Robert S.H. Istepanian and James F. Whidborne (Eds.)

Optimisation of Industrial Processes at Supervisory Level
Doris Sáez, Aldo Cipriano and Andrzej W. Ordys

Applied Predictive Control
Huang Sunan, Tan Kok Kiong and Lee Tong Heng

Hard Disk Drive Servo Systems
Ben M. Chen, Tong H. Lee and Venkatakrishnan Venkataramanan

Robust Control of Diesel Ship Propulsion
Nikolaos Xiros

Model-based Fault Diagnosis in Dynamic Systems Using Identification Techniques
Silvio Simani, Cesare Fantuzzi and Ron J. Patton

Strategies for Feedback Linearisation
Freddy Garces, Victor M. Becerra, Chandrasekhar Kambhampati and
Kevin Warwick

Robust Autonomous Guidance
Alberto Isidori, Lorenzo Marconi and Andrea Serrani

Dynamic Modelling of Gas Turbines
Gennady G. Kulikov and Haydn A. Thompson (Eds.)

Fuzzy Logic, Identification and Predictive Control
Jairo Espinosa, Joos Vandewalle and Vincent Wertz

Optimal Real-time Control of Sewer Networks
Magdalene Marinaki and Markos Papageorgiou
Publication due December 2004

Adaptive Voltage Control in Power Systems
Giuseppe Fusco and Mario Russo
Publication due August 2005

Jay T. Pukrushpan, Anna G. Stefanopoulou
and Huei Peng

Control of Fuel Cell Power Systems

Principles, Modeling, Analysis and Feedback Design

With 111 Figures

 Springer

Jay T. Pukrushpan, PhD
Department of Mechanical Engineering, Kasetsart University, Bangkok, Thailand

Anna G. Stefanopoulou, PhD
Huei Peng, PhD
Department of Mechanical Engineering, University of Michigan, Ann Arbor, MI 48109, USA

British Library Cataloguing in Publication Data
Pukrushpan, Jay T.
    Control of fuel cell power systems : principles, modeling,
    analysis and feedback design. — (Advances in industrial
    control)
    1. Fuel cells 2. Electric power systems — Control
    3. Automobiles — Motors — Control systems 4. Automobiles —
    Fuel systems
    I. Title II. Stefanopoulou, Anna G. III. Peng, Huei
    629.2'53
ISBN 1852338164

Library of Congress Cataloging-in-Publication Data
Pukrushpan, Jay T.
    Control of fuel cell power systems : principles, modeling, analysis, and feedback design /
Jay T. Pukrushpan, Anna G. Stefanopoulou, Huei Peng.
        p. cm — (Advances in industrial control)
    Includes bibliographical references and index.
    ISBN 1-85233-816-4 (hc : alk. paper)
        1. Fuel cells. 2. Control theory. I. Stefanopoulou, Anna G. II. Peng, Huei.
    III. Title. IV. Series.
    TK2931.P85   2004
    621.31'2429—dc22   2004049905

ISBN 1-85233-816-4
Springer Science+Business Media
springeronline.com

Typesetting: Electronic text files prepared by author
Printed in the United States of America
69/3830-54321   Printed on acid-free paper   SPIN 11417637

Advances in Industrial Control

Series Editors

Professor Michael J. Grimble, Professor Emeritus of Industrial Systems and Director
Professor Michael A. Johnson, Professor of Control Systems and Deputy Director

Industrial Control Centre
Department of Electronic and Electrical Engineering
University of Strathclyde
Graham Hills Building
50 George Street
Glasgow G1 1QE
United Kingdom

Professor Emeritus O.P. Malik
Department of Electrical and Computer Engineering
University of Calgary
2500, University Drive, NW
Calgary
Alberta
T2N 1N4
Canada

Professor K.-F. Man
Electronic Engineering Department
City University of Hong Kong
Tat Chee Avenue
Kowloon
Hong Kong

Professor G. Olsson
Department of Industrial Electrical Engineering and Automation
Lund Institute of Technology
Box 118
S-221 00 Lund
Sweden

Professor A. Ray
Pennsylvania State University
Department of Mechanical Engineering
0329 Reber Building
University Park
PA 16802
USA

Professor D.E. Seborg
Chemical Engineering
3335 Engineering II
University of California Santa Barbara
Santa Barbara
CA 93106
USA

Doctor I. Yamamoto
Technical Headquarters
Nagasaki Research & Development Center
Mitsubishi Heavy Industries Ltd
5-717-1, Fukahori-Machi
Nagasaki 851-0392
Japan

Series Editors' Foreword

The series *Advances in Industrial Control* aims to report and encourage technology transfer in control engineering. The rapid development of control technology has an impact on all areas of the control discipline. New theory, new controllers, actuators, sensors, new industrial processes, computer methods, new applications, new philosophies ... , new challenges. Much of this development work resides in industrial reports, feasibility study papers and the reports of advanced collaborative projects. The series offers an opportunity for researchers to present an extended exposition of such new work in all aspects of industrial control for wider and rapid dissemination.

Fuel cell power systems is a new and exciting industrial area which is receiving considerable commercial investment as a future energy technology. The IEEE *Spectrum* magazine New Year issue of 2004 cited this area as one whose development and progress should be observed closely over the coming years. Successful application of fuel cell technology will depend on many factors not the least of which being how consistently their performance can be controlled. This *Advances in Industrial Control* monograph by Jay Pukrushpan, Anna Stefanopoulou and Huei Peng is a timely contribution to the area. As the authors so rightly say in their Preface, control engineers have different requirements from modelling, experimental studies and simulation work for designing a good control system and this monograph presents a *control-orientated* approach to these topics in the fuel cell power system field.

For the fuel cell and control specialist a good few chapters in this monograph are devoted to various system models; a further indication of the crucial importance of modelling, simulation and basic system comprehension to enable control engineers to develop a well designed control system. A fully developed control system design exercise appears in Chapter 7 of the monograph. As well as being of interest to fuel cell engineers, the chapter is of pedagogical interest to the general control engineer. The control system to be designed is multivariable and the steps; input-output pairing, decentralized control and then multivariable control design, are followed. Of particular interest is the insight obtained from the multivariable control designs which is then used to suggest improvements to the decentralized control design. The final chapters in the book provide a useful look at future research directions.

The monograph is a very interesting addition to the *Advances in Indus^trial Control* series and a valuable contribution to the growing engineering area of fuel cell power system technology.

M.J. Grimble and M.A. Johnson
Industrial Control Centre
Glasgow, Scotland, U.K.

Preface

Fuel cell systems offer clean and efficient energy production and are currently under intensive development by several manufacturers for both stationary and mobile applications. The viability, efficiency, and robustness of the fuel cell technology depend on understanding, predicting, monitoring, and controlling the fuel cell system under a variety of environmental conditions and a wide operating range.

Many publications have discussed the importance and the need for a well-designed control system for fuel cell power plants. From discussions with control engineers and researchers in the area of fuel cell technology it became apparent that a comprehensive book with a control-oriented approach to modeling, analysis, and design was needed. The field is fast evolving and there is a lot of excitement but also a lot of commercial or confidentiality considerations that do not allow state-of-the-art results to be published.

In this book, we address this need by developing phenomenological models and applying model-based control techniques in polymer electrolyte membrane fuel cell systems. The book includes:

- An overview and comprehensive literature survey of polymer electrolyte membrane fuel cell systems, the underlying physical principles, the main control objectives, and the fundamental control difficulties.
- System-level dynamic models from physics-based component models using flow characteristics, point-mass inertia dynamics, lumped-volume manifold filling dynamics, time-evolving spatially homogeneous reactant pressure or concentration, and simple diffusion, transport, and heat equations.
- Formulation, in-depth analysis, and detailed control design for two critical control problems, namely, the control of the cathode oxygen supply for a high-pressure direct hydrogen Fuel Cell System (FCS) and control of the anode hydrogen supply from a natural gas Fuel Processor System (FPS).
- Multivariable controllers that address subsystem conflicts and constraints from sensor fidelity or actuator authority.

- Real-time observers for stack variables that may be hard to measure or augment existing stack sensors for redundancy in fault detection.
- Examples where control analysis not only can be used to develop robust controllers but also can help in making decisions on fuel cell system redesign for improved performance.
- More than 100 figures and illustrations.

This book is intended for researchers and students with basic control knowledge but who are novices in fuel cell technology. The simplicity of the models and the application of the control algorithms in concrete case studies should help practicing fuel cell engineers. Other scientists from electrochemistry, material sciences, and fluid dynamics who wish to become familiar with the control tools and methods may also benefit from the comprehensive coverage of the control design. Managers or entrepreneurs interested in accessing the challenges and opportunities in fuel cell automation technology may also find this book useful.

Book Overview

The development of a model of a dynamic fuel cell reactant supply subsystem that is suitable for control study is explained in Chapters 2 and 3. The model incorporates the transient behaviors that are important for integrated control design and analysis. Models of the auxiliary components, namely, a compressor, manifolds, an air cooler, and a humidifier, are presented in Chapter 2. Inertia dynamics along with nonlinear curve fitting of the compressor characteristic map are used to model the compressor. The manifold dynamic models are based on lumped-volume filling dynamics. Static models of the air cooler and air humidifier are developed using thermodynamics.

The fuel cell stack model in Chapter 3 is composed of four interacting submodels, namely, stack voltage, cathode flow, anode flow, and membrane hydration models. The stack voltage is calculated as a function of stack current, cell temperature, air pressure, oxygen and hydrogen partial pressures, and membrane humidity. The voltage function presented in Section 3.1 is based on the Nernst open circuit voltage, and activation, ohmic, and concentration losses. Flow equations, mass continuity, and electrochemical relations are used to create lumped-parameter dynamic models of the flow in the cathode and anode in Sections 3.2 and 3.3. Mass transport of water across the fuel cell membrane is calculated in the membrane hydration model in Section 3.4.

A perfect controller for air humidification and a simple proportional controller of the hydrogen supply valve are integrated into the model to allow us to focus on the analysis and control design of the air supply system. In Chapter 4, we perform a steady-state analysis of the model in order to determine the optimal value of the air flow setpoint, termed oxygen excess ratio, that results in the maximum system net power. The resulting value agrees with the fuel cell specification in the literature, and thus indirectly validates the accuracy of the model. Results from the simulation of the model with a static feedforward controller based on the optimal setpoint are presented in Section 4.3. The model predicts transient behavior similar to that reported in the literature.

The control design of the air supply system using model-based linear control techniques is presented in Chapter 5. The goal of the control problem is to effectively regulate the oxygen concentration in the cathode by quickly and accurately replenishing oxygen depleted during power generation. Several control configurations are studied and the advantages and disadvantages of each configuration are also explained. Additionally, the performance limitations of the controller due to measurement constraints are also illustrated. In Section 5.5.2, the results from an observability analysis suggest the use of stack voltage measurement in the feedback to improve the performance of the observer-based controller. The analogy between the fuel cell closed-loop current-to-voltage transfer function and an electrical impedance, discussed in Section 5.6, can be useful to researchers in the area of power electronics. Section 5.7 presents an analysis of the tradeoff between regulation of cathode oxygen and desired net power during transient. A range of frequencies associated with the tradeoff is determined.

In Chapters 6 and 7, a control problem of the partial oxidation based natural gas fuel processor is studied. The components and processes associated with the processor are explained in Section 6.1. A dynamic model of the processor is also presented in Chapter 6. Transient flow, pressure, and reactor temperature characteristics are included. The reaction products are determined based on the chemical reactions, and the effects of both the oxygen-to-carbon ratio and the reactor temperature on the conversion are included. The model is validated with a high-order detailed model of the fuel cell and fuel processor system, and the results are shown in Section 6.3.

A two-input two-output control problem of regulating the catalytic partial oxidation (CPOX) temperature and the stack anode hydrogen concentration using natural gas valve and air blower commands is studied in Chapter 7. Section 7.3 illustrates the use of the relative gain array method to find appropriate pairings of the system input and output and also to analyze the system interactions. The analysis shows that large system interactions degrade the performance of the decentralized controller, especially during transient operation. A model-based multivariable controller for the fuel processor system is designed in Section 7.5 using the linear quadratic optimal method. It is shown that significant improvement in CPOX temperature regulation can be achieved with the designed multivariable controller. The controller is then analyzed to determine the important terms that contribute to the improvement of the closed loop performance. This will be useful in the simplification and implementation of the controller. Chapter 8 provides a summary and contributions of the work. Several topics that need to be addressed and several other interesting areas to study are also given.

The major technical topics covered in this book are:

- Two control problems of the fuel cell power generation system are formulated. The first problem is the control of the air supply system for a high-pressure direct hydrogen fuel cell system (FCS). The objective is to control the compressor motor command to quickly and efficiently replenish the cathode oxygen depleted during system power generations. The second problem is the control of a low-pressure natural gas fuel processor system (FPS). The goal is to coordinate an air blower and a fuel valve in order to quickly replenish the hydrogen depleted in the fuel cell anode while maintaining the desired temperature of the catalytic partial oxidation reactor.
- Control-oriented dynamic models suitable for control design and analysis are created. The complexity of the models is kept minimal by considering only physical effects relevant to the control problems. The models are developed using physics-based principles allowing them to be used for different fuel cell systems requiring only parameter modifications. Moreover, the variables in the models represent real physical variables providing insight into the dynamic behavior of the real system. The causality of the process is clearly demonstrated in the models.
- The models are used in the model-based control analysis to develop controllers and to determine required control structures that provide an enhanced performance over conventional controllers. Moreover, the analysis provides insight into the performance limitations associated with plant architecture, sensor location, and actuator bandwidth.
  - For the FCS, the limitations of using integral control and an observer-based controller arise from sensor locations. In particular, a direct measurement of the performance variable (*i.e.*, the oxygen excess ratio) is not possible. The compressor flow rate, which is located upstream from the stack, is traditionally used as the only feedback to the controller. Our observability analysis shows that the stack voltage measurement can be used to enhance the closed-loop system performance and robustness. The voltage measurement is currently used only for safety monitoring. However, we demonstrate that the fuel cell stack mean voltage can be used for active control of fuel cell stack starvation. This result exemplifies the power of control-theoretic tools in defining critical and cost-effective sensor location for the FCS.
  - An additional limitation arises when the FCS architecture dictates that all auxiliary equipment is powered directly from the fuel cell with no secondary power sources. This plant configuration is preferred due to its simplicity, compactness, and low cost. We used linear optimal control design to identify the frequencies at which there is severe tradeoff between the transient system net power performance and the stack starvation control. The result can be used to determine the required size of additional energy or oxygen storage devices in the case where fast transient response is required. We demonstrated that the multi-

variable controller improves the performance of the FCS and results in a different current–voltage dynamic relationship that is captured by the closed-loop FCS impedance. We expect that the derived closed-loop FCS impedance will be very useful and will provide the basis for a systematic design of fuel cell electronic components.

- Multivariable feedback analysis using the control-oriented model of the FPS indicates large system interactions between the fuel and the air loops at high frequencies. Our analysis shows that the magnitude and speed of the fuel valve limit the closed-loop bandwidth in the fuel loop, and thus affect hydrogen starvation. We demonstrate that fast regulation of CPOX temperature, which is the objective in the air loop, requires a fast blower and air dynamics if a decentralized control structure is used. On the other hand, a slow blower can also accomplish similar performance if it is coordinated with the fuel valve command. The coordination is achieved with a model-based controller that decouples the two loops at the frequencies of high interaction. With this result we provide rigorous guidelines regarding actuator specifications and the necessary software complexity for multiple actuator coordination.

Acknowledgments

We would like to the acknowledge the support of the Royal Thai government, the Automotive Research Center at the University of Michigan, and the National Science Foundation for their financial support during the period in which this manuscript was written.

We wish to thank Subbarao Varigonda, Jonas Eborn, Thordur Runolfsson, Christoph Haugstetter, Lars Pedersen, Shubhro Ghosh, Scott Bortoff, and Clas A. Jacobson from the United Technology Research Center for their help and for sharing with us their knowledge of fuel processor systems. We are grateful to Scott Staley, Doug Bell, Woong-Chul Yang, and James Adams of the Ford Motor Company for their encouragement and for providing valuable data on a vehicle fuel cell system.

At the University of Michigan, in the vehicle control laboratory and in the powertrain control laboratory we have worked alongside many talented students. Our thanks go to all of them. Thanks are also due to Dr. James Freudenberg and Dr. Erdogan Gulari from the University of Michigan for their help and advice.

Bangkok, Thailand   *Jay T. Pukrushpan*
Ann Arbor, Michigan   *Anna G. Stefanopoulou*
Ann Arbor, Michigan   *Huei Peng*
January 2004

Contents

1

Background and Introduction

Fuel cells are electrochemical devices that convert the chemical energy of a gaseous fuel directly into electricity and are widely regarded as a potential alternative stationary and mobile power source. They complement heat engines and reduce the ubiquitous dependence on fossil fuels and thus have significant environmental and national security implications. As such, they are actively studied for commercial stationary power generation, residential applications, and transportation technologies. Recent study has shown that, in the United States, carbon dioxide (CO_2) accounts for more than 80% of greenhouse gases released [117] and the transportation sector is responsible for 32% of the overall CO_2 emission [31]. In this book, we concentrate on the fuel cell control requirement during transients. Application of fuel cells in automotive powertrains is emphasized, partly because ground vehicle propulsion conditions present the most challenging control problem, and partly due to their importance in global fuel consumption and emission generation.

Fuel cell stack systems are under intensive development by several manufacturers, with the Polymer Electrolyte Membrane (PEM) fuel cells (also known as Proton Exchange Membrane fuel cells) currently considered by many to be in a relatively more developed stage for ground vehicle applications. PEM fuel cells have high power density, solid electrolyte, long cell and stack life, as well as low corrosion. They have greater efficiency when compared to heat engines and their use in modular electricity generation and propulsion of electric vehicles is promising [62]. Fuel cell efficiency is high at partial load which corresponds to the majority of urban and highway driving scenarios [90]. At a nominal driving speed (30 mph) the efficiency of a fuel cell electric drive using direct hydrogen from natural gas is two times higher than that of a conventional internal combustion engine [94]. Using pure hydrogen as fuel can eliminate local emissions problems in densely populated urban environments. A hydrogen generation and distribution infrastructure based on renewable energy from wind, water, and solar energy, or fuel processors will help reduce our dependency on fossil fuels.

To compete with existing Internal Combustion Engines (ICE), however, fuel cell systems must operate at similar levels of performance. Transient behavior is one of the key requirements for the success of fuel cell vehicles. Efficient fuel cell system power response depends on the air and hydrogen feed, flow and pressure regulation, and heat and water management. As current is instantaneously drawn from the load source connected to the fuel cell stack, heat and water are generated, whereas oxygen is depleted. During this transient, the fuel cell stack breathing control system is required to maintain proper temperature, membrane hydration, and partial pressure of the reactants across the membrane to avoid detrimental degradation of the stack voltage, and thus efficiency reduction. These critical fuel cell parameters can be controlled for a wide range of currents, and thus power, by a series of actuators such as valves, pumps, compressor motors, expander vanes, fan motors, humidifiers, and condensers. The resulting auxiliary actuator system is needed to make fine and fast adjustments to satisfy performance, safety, and reliability standards that are independent of age and operating conditions. Model-based dynamic analysis and control design give insight into the subsystem interactions and control design limitations. They also provide guidelines for sensor selection and control coordination between subsystems. Creating a control-oriented dynamic model of the overall system is an essential first step not only for understanding system behavior but also for the development and design of the model-based control methodologies. This book presents first the development of physics-based dynamic models of fuel cell systems and fuel processor systems and then the applications of multivariable control techniques to study their behavior. The analysis gives insight into the control design limitations and provides guidelines for the necessary controller structure and system redesign.

1.1 Fuel Cell

We summarize here the principle and potential benefits of fuel cell power generation. The fuel cell principle was discovered in 1839 by William R. Grove, a British physicist [54]. A fuel cell consists of an electrolyte sandwiched between two electrodes. The electrolyte has a special property that allows positive ions (protons) to pass through while blocking electrons. Hydrogen gas passes over one electrode, called an anode, and with the help of a catalyst, separates into electrons and hydrogen protons (Figure 1.1),

$$2\mathrm{H}_2 \Rightarrow 4\mathrm{H}^+ + 4e^- \tag{1.1}$$

The protons flow to the other electrode, called a cathode, through the electrolyte while the electrons flow through an external circuit, thus creating electricity. The hydrogen protons and electrons combine with oxygen flow through the cathode, and produce water.

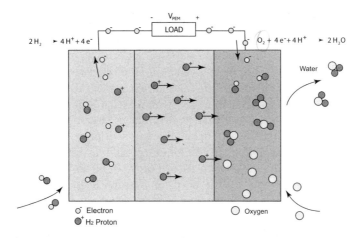

Fig. 1.1. Fuel cell reaction

$$O_2 + 4H^+ + 4e^- \Rightarrow 2H_2O \qquad (1.2)$$

The overall reaction of the fuel cell is therefore

$$2H_2 + O_2 \Rightarrow 2H_2O \qquad (1.3)$$

The voltage produced from one cell is between 0 and 1 volts [69] depending on fuel cell operating conditions and the size of the load connected to the fuel cell. The typical value of the fuel cell voltage is about 0.7 volts. To get higher voltage, multiple cells are stacked in series. The total stack voltage is the number of cells multiplied by the average cell voltage. As with other electrical devices, there are electrical resistances in the fuel cell. The loss associated with the resistance is dissipated in the form of heat. In other words, heat is released from the fuel cell reaction.

Fuel cells have several advantages over internal combustion engines and batteries. To generate mechanical energy, the ICE first converts fuel energy to thermal energy by combusting fuel with oxygen at high temperature. The thermal energy is then used to generate mechanical energy. Because thermal energy is involved, the efficiency of the conversion process is limited by the Carnot Cycle [112]. Unlike ICE, fuel cells directly convert fuel energy to electrical energy and its maximum efficiency is not subjected to Carnot Cycle limitations. Higher energy conversion efficiency can potentially be achieved by fuel cells. If hydrogen is used as fuel, the outcome of the fuel cell reaction is water and heat. Therefore, fuel cells are considered to be a zero emission power generator. They do not create pollutants such as hydrocarbon or nitrogen oxide. A battery is also an electrochemical device that converts chemical energy directly to electricity. However, the battery reactants are stored internally and when used up, the battery must be recharged or replaced. The reactants of the fuel cell are stored externally. Oxygen is typically taken from atmospheric air and hydrogen is stored in high-pressure or cryogenic tanks which can be refueled. Refilling fuel tanks requires significantly less time than recharging batteries [112].

There are different types of fuel cells, distinguished mainly by the type of electrolyte used. The differences in cell characteristics, such as cell material, operating temperature, and fuel diversity, make each type of fuel cell suitable for different applications. It is known that Polymer Electrolyte Membrane Fuel Cells (PEMFC) are suitable for automobile applications. PEM fuel cells have high power density, a solid electrolyte, and long life, as well as low corrosion. PEM fuel cells operate in the temperature range of 50 to 100°C which allows safer operation and eliminates the need for thermal insulation. The polymer electrolyte membrane is an electronic insulator but an excellent conductor of hydrogen ions. The typical membrane material consists of a fluorocarbon backbone to which sulfonic acid groups ($SO_3^-H^+$) are attached [112]. When the membrane becomes hydrated, the hydrogen ions (H^+) in the sulfonic group are mobile. Depending on membrane manufacturers and the versions of the membrane, properties of the membranes differ. The thickness of the membrane varies from 50 to 175 microns, which is approximately 2 to 7 papers thick [112]. The membrane is sandwiched between two electrodes (anode and cathode) made from a highly conducting material such as porous graphite. A small amount of platinum is applied to the surface of the anode and cathode to help increase the rate of reaction. The three components (anode, electrolyte, and cathode) are sealed together to form a single membrane electrolyte assembly (MEA), shown in Figure 1.2, which is typically less than a millimeter thick.

The MEA is sandwiched by two backing layers made from porous carbon. The porous nature of the backing layer ensures effective diffusion of each reactant gas to the catalyst site on the MEA. The outer surface of the backing layer is pressed against the flow field plates which serve as both reactant gas flow field and current collector. The plate is made of a lightweight, strong, gas impermeable, electron conducting material such as graphite or composite materials. The other side of the flow field plate is connected to the next cell. The number of cells stacked in one fuel cell stack depends on the power requirement of the stack, which varies across different applications.

Typical characteristics of fuel cells are normally given in the form of a polarization curve, shown in Figure 1.3, which is a plot of cell voltage versus cell current density (current per unit cell active area). The differences between actual voltage and the ideal voltage of the fuel cell represent the loss in the cell. As shown in Figure 1.3, as more current is drawn from the fuel cell, the voltage decreases, due to fuel cell electrical resistance, inefficient reactant gas transport, and low reaction rate. Because lower voltage indicates lower efficiency of the fuel cell, low load (low current) operation is preferred. However, this will increase the fuel cell volume and weight. Moreover, constant operation at low load is not practical in automobile applications where frequent load changes are demanded. The polarization curve shown in Figure 1.3 is for a specific operating condition. The curve varies with different operating con-

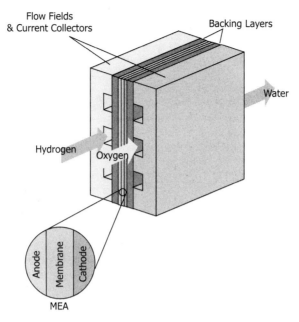

Fig. 1.2. Fuel cell structure

ditions, including different pressure, temperature, reactant partial pressure, and membrane humidity. An example of pressure effects on the polarization curve is shown in Figure 1.4. The data, kindly given to us by the Ford Research Laboratory, are from a generic PEM fuel cell stack used in a fuel cell prototype vehicle.

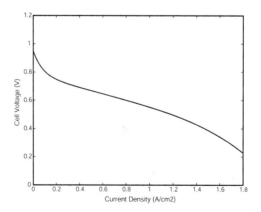

Fig. 1.3. Typical fuel cell polarization curve

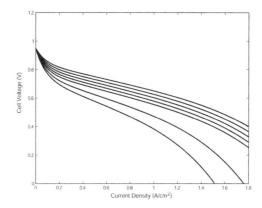

Fig. 1.4. Fuel cell polarization for different operating pressures

1.2 Fuel Cell Propulsion System for Automobiles

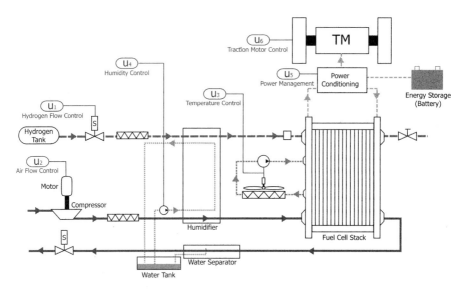

Fig. 1.5. Automotive fuel cell propulsion system

A fuel cell stack needs to be integrated with several auxiliary components to form a complete fuel cell system. The diagram in Figure 1.5 shows the minimal components required for a pressurized fuel cell system. The fuel cell stack requires four flow systems: hydrogen supply system to the anode; air supply system to the cathode; de-ionized water serving as coolant in the stack cooling channel; and de-ionized water supply to the humidifier to humidify the hydrogen and the air flows. These four subsystems are denoted by control inputs u_1 to u_4 in Figure 1.5.

Operation at high pressure significantly improves the reaction rate and, thus, the fuel cell efficiency and power density [8]; a compressor and an electric drive motor are needed not only to achieve the desired air flow but also to compress air to a desired pressure level. The pressurized air flow leaving the compressor is at a higher temperature, therefore an air cooler is needed to reduce the air temperature before it enters the stack. A humidifier is used to add vapor into the air flow, as illustrated in Figure 1.5, in order to prevent dehydration of the membrane. As the air leaves the stack, it carries vapor produced in the fuel cell. For an automotive fuel cell system, a water separator is needed in order to recover the water to be used to humidify the reactants. On the anode side, hydrogen is supplied from a container that can store pressurized hydrogen or liquid hydrogen. A valve is used to control the flow rate of hydrogen. The hydrogen flow is, in some systems, humidified before entering the stack. Hydrogen and air react in the stack producing electricity, water, and heat. Because the temperature of the stack must be maintained below 100°C for the membrane to be properly humidified, excessive heat released in the fuel cell reaction is removed by a de-ionized water coolant. As illustrated in Figure 1.5, the coolant leaving the stack then passes through a heat exchanger or a radiator in order to remove heat from the system. A power conditioner, denoted as control input u_5 in Figure 1.5, is usually needed because the voltage of the fuel cell stack varies significantly, which is not suitable for typical electronic components nor traction motors. The conditioned power is supplied to the traction motor connected to the vehicle drivetrain. The control input u_6 in Figure 1.5 represents the control of the traction motor drive.

Reactant flow rate, total pressure, reactant partial pressure, temperature, and membrane humidity are the main parameters that need to be regulated in order to ensure (i) fast system transient response, consistent warm-ups, and safe shut-down, and (ii) system robustness and adaptation to changing power. The main control devices are the compressor motor for the air flow and pressure regulations, the valve for hydrogen flow rate and pressure regulations, the water pump or radiator fan speed for the temperature regulations, and the humidifier for the humidity control. However, the changes in the parameters are not independent. Changes in one parameter influence the others. For example, an increase in air flow rate can cause an increase in air pressure but can also vary the amount of vapor and heat entering and leaving the stack, thus affecting the humidity of the membrane and temperature of the stack. Stack temperature also affects the humidity of the air and hydrogen inside the stack because the vapor saturation pressure depends strongly on the temperature.

During vehicle operation, various load levels as well as sudden load changes are expected. For fuel cell vehicles to be commercialized, these vehicle operations need to be well handled. During this transient, the control system is required to maintain optimal temperature, membrane hydration, and partial pressure of the reactants in order to avoid detrimental degradation of the fuel cell voltage and, thus, an efficiency reduction and fuel cell life shortening.

1.3 System Interactions

Precise control of the reactant flow and pressure, stack temperature, and membrane humidity is critical to the viability, efficiency, and robustness of fuel cell propulsion systems. The resulting task is complex because of subsystem interactions and conflicting objectives. The overall system could be partitioned into four subsystems. Each system has a corresponding control objective and also interactions with other subsystems. The subsystems are the reactant flow, the heat and temperature, the water management, and the power management subsystems.

1.3.1 Reactant Flow Subsystem

The reactant flow subsystem consists of hydrogen supply and air supply loops. As the vehicle traction motor draws current, hydrogen and oxygen become depleted in the fuel cell stack. In the case where compressed hydrogen is available on-board the hydrogen flow in the anode and the air flow in the cathode are adjusted using a valve and a positive-pressure flow device, respectively. The control objective is to provide sufficient reactant flows (to keep the desired excess ratio) to ensure fast and safe power transient responses and to minimize auxiliary power consumption. A high-pressure fuel cell system that directly drives its own compressor reacts to a positive step in requested power with an inverse response of the power delivered. Although the inverse response can be avoided by using a battery-driven motor, the direct coupling of the fuel cell with the compressor is the preferred configuration due to efficiency benefits and compactness. The nonminimum phase behavior of the fuel cell power output limits the closed-loop bandwidth of this loop. A few early patents [75, 82] recognize this difficulty and avoid a slow response by relying on a feedforward map that must be tuned at different ambient conditions [85]. Several experimental systems use a fixed speed motor which supplies air flow that satisfies maximum traction requirements. This results in unnecessary auxiliary power consumption during low-load operations where less flow is needed.

In a low-pressure fuel cell system a low-speed blower is utilized for supplying the air. The blower requires less power than the compressor, and consequently, alleviates the FCS inverse response. The blower inertia becomes then the limiting factor for the speed of the FCS power response. Other consequences of employing a low-pressure FCS, such as the humidification requirements and the FCS power density, are currently under investigation.

1.3.2 Heat and Temperature Subsystem

The heat and temperature subsystem includes the fuel cell stack cooling system and the reactant temperature system. As current is drawn by the traction motor, heat is generated in the fuel cell. With the stack size required for passenger vehicles, the heat generated cannot be passively dissipated by air convection and radiation through the external surface of the stack. This requires active cooling through the reactant flow rate and the cooling system. The thermal management of the fuel cell stack is more challenging than that of the internal combustion engine. First, de-ionized water is used as the coolant in the stack instead of an effective coolant fluid. Second, the PEM fuel cell is designed to operate at the temperature around 80°C. Therefore, the exhaust air exiting the stack, which has temperature around 80°C, has less ability to carry out heat than the ICE exhaust gas which is over 500°C [58]. Heat rejection for the fuel cell stack is therefore a responsibility of the cooling system. Furthermore, the low temperature difference between the stack and the water coolant limits the effectiveness of the heat transfer from the stack to the coolant. Apart from the water coolant flow rate and its temperature, the temperature of inlet reactant air also affects the temperature of the stack. The heat management system can vary the speed of the cooling fan and the recirculation pump in coordination with adjusting a bypass valve. The goal of thermal management is fast warm-up with no stack temperature overshoot and low auxiliary fan and pump power.

1.3.3 Water Management Subsystem

The task of the water management system is to maintain hydration of the polymer membrane and to balance water usage/consumption in the system. The amount of reactant flow and the water injected into the anode and cathode flow streams affect the humidity of the membrane. Dry membranes and flooded fuel cells cause high polarization losses. As the current is drawn from the fuel cell, water molecules are both produced in the cathode and dragged from the anode to the cathode by the hydrogen protons. As the concentration of water in the cathode increases, the concentration gradient causes water to diffuse from the cathode to the anode. Perturbation in fuel cell humidity can be caused by different mechanisms: water generated while load increases, changes in the absolute and relative reactant pressure across the membrane, changes in air flow rate, and changes in stack temperature, which change the vapor saturation pressure. These mechanisms indicate strong and nonlinear interactions among the humidity control tasks, the reactant flow management loop, the heat management loop, and the power management loop. A 20 to 40% drop in voltage can occur if there is no proper humidification control [24].

1.3.4 Power Management Subsystem

The power management subsystem controls the power drawn from the fuel cell stack. Without considering power management, the load current can be viewed as a disturbance to the fuel cell system. However, as shown above, the drawn current has a direct impact on other subsystems. If a battery is used as another power source in the system, the power management between two power sources could be applied with the objective of giving a satisfactory vehicle transient response, achieving optimal system efficiency, and assisting the fuel cell system.

1.3.5 Fuel Processor Subsystem

Inadequate infrastructure for hydrogen refueling, distribution, and storage makes fuel processor technology an important part of the fuel cell system. Methanol, gasoline, and natural gas are examples of fuels being considered as fuel cell energy sources. Figure 1.6 illustrates different processes involved in converting carbon-based fuel to hydrogen [18, 23]. Interactions between

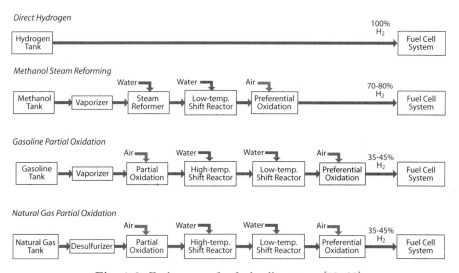

Fig. 1.6. Fuel sources for fuel cell systems [18, 23]

the components and many additional control actuators in the fuel processor introduce additional complexity to the control problem. In addition to the fuel cell variables, the fuel processor variables that require precise control include the temperature of the reactors and the concentration of hydrogen and carbon monoxide in the gas stream.

1.4 Literature Review

To achieve high efficiency and a long lifecycle of the fuel cell stack, the reactant gas supply and water and heat subsystems need to be properly controlled both during steady-state and transient operations. During vehicle transient operation, delivering torque to meet driveability performance while meeting safety and efficiency criteria is of concern [107]. Yang et al. [120] described the control challenges and the methodologies being used in fuel cell prototype vehicles. A variety of control problems was identified and discussed. Another report [64] discussed the importance of subsystem management, or balance of plant, and control needed for each subsystem. The difficulties of the thermal management system were explained in [46]. The interactions between thermal management and stack performance were also addressed. Several integration issues and tradeoffs within the fuel cell system were discussed. The impact of stack water and thermal managements on the fuel cell system was studied in [11]. In [13] and [47], fuel cell stack in-vehicle performance was shown to be lower than performance on a laboratory test-stand due to the discrepancy in fuel cell operating conditions, particularly inadequate air supply and insufficient humidification. The need for control strategies that can respond fast and can regulate the fuel cell operating conditions was emphasized.

Despite a large number of publications on fuel cell modeling, models of fuel cell systems suitable for control studies are still lacking. The models developed in the literature can be classified into three main categories, namely, detailed fuel cell models based on partial differential equations, steady-state fuel cell system models based on experimental maps or look-up tables, and dynamic fuel cell system models that neglect spatial variations.

Most of the publications on fuel cell modeling were developed at the cell level and included spatial variations of the fuel cell parameters. Complex electrochemical, thermodynamic, and fluid mechanics principles were used to develop these models. The performance or efficiency of the fuel cell under different steady-state operating conditions can be determined using this type of model. The main purposes of these models are to design the fuel cell components and to choose the fuel cell operating points. Although these models are not suitable for control studies, they establish the fundamental effects of operating parameters, such as pressure and temperature, on the fuel cell voltage. Several publications [4, 65, 77, 78] presented the formulation of fuel cell resistances which is used to predict fuel cell polarization characteristics at different operating conditions. Mass transport of gas and water was also included in several publications with both one-dimensional [5, 16, 17, 104] and two-dimensional models [30, 55, 100]. Springer et al. [104] also presented a model predicting net water flow per proton through the membrane and the increase in membrane resistance due to the membrane water content. Many publications addressed the water and thermal management of the fuel cell. Nguyen and White [83] developed a model investigating the effectiveness of various humidification designs. Fuller and Newman [48] developed a two-dimensional mass transport model of a membrane electrode assembly to examine the water, thermal, and reactant utilization of the fuel cell. Okada *et al.* [87] presented a method to analytically calculate water concentration profiles in the membrane. A three-dimensional numerical model that predicts the mass flow between the cathode and anode channels was presented in [39]. Bernardi [15] and Büchi and Srinivasan [24] presented models that identify operating conditions that result in water balance in the fuel cell. Baschuk and Li [14] developed a model that includes the effect of water flooding in the cathode catalyst layer. Wöhr *et al.* [119] presented a dynamic model of heat and water transport in the fuel cell and showed the effects of various current density variations on the fuel cell performance. Interestingly, they showed that different rates of load changes can lead to a different level of fuel cell voltage as a result of water deficiency. Several models were developed to represent fuel cell stacks [72, 110]. In [72], the model was used to determine operating configurations. The stack model in [110] was used in the stack flow field design. A model predicting transient responses of the fuel cell stack was given in [6]. The heat transfer transient phenomena were incorporated into this model. All the papers in the above category used a combination of experiments and physical laws to derive their models.

An interesting set of papers with experimental results of fuel cell performance during dynamic excitation appeared in the literature recently. Specifically, Chu and Jiang evaluated fuel cell performance under various conditions. Different types of membrane were tested in [27] and the humidity and hydrogen flow effects were presented in [28]. The voltage-time behaviors of the fuel cell stack at constant current discharge were studied and a model representing the behavior was presented in [61]. The stack structure designs were tested in [60]. Laurencelle *et al.* [70] presented experimental results of fuel cell stack responses during load transitions. The transient behavior of stack voltage during positive load switching was observed in the experiment.

Steady-state system models are typically used for component sizing, static tradeoff analysis, and cumulative fuel consumption or hybridization studies. The models in this category represent each component such as the compressors, heat exchangers, and fuel cell stack voltage as a static performance or efficiency map. The only dynamics considered in this type of model is the vehicle effective inertia. Barbir *et al.* [12] presented a steady-state model of the entire system that calculates the system and component parameters for various operating pressures, temperatures, and power levels. System efficiency was also evaluated. The size of the heat exchanger or radiator was determined for each system configuration. Equations presented in [45] were used to find operating strategies based on the efficiency of each individual component in an indirect methanol fuel cell system. A method to optimize the net power output was presented. The fuel cell system models in [3, 10, 19, 84] were used in fuel cell/battery hybrid studies. Fuel economy was determined and supervisory vehicle control was studied using the model in [19]. The model in [3] was used to study the tradeoff between maximum acceleration and auxiliary power sources. The vehicle inertia dynamics were the only transient phenomena in this model. Sizing of the fuel cell and battery in a hybrid configuration was studied in [10]. This model was used to choose the degree of hybridization that offers high fuel economy and to study power management strategies between the fuel cell stack and battery. Steady-state models of the fuel cell stack, air supply system, and thermal management system were incorporated into a vehicle simulation program in [98]. The model was used to predict the acceleration, braking, and drive cycle fuel economy performance of a fuel cell stack and ultracapacitor hybrid SUV vehicles. In most of these papers, the fuel cell stack was modeled with a static polarization relationship for fixed fuel cell operating parameters.

Several dynamic fuel cell system models exist in the literature. Different levels of dynamic behavior were incorporated into each of the models. The thermal dynamics are considered to be the slowest dynamics in the fuel cell system. Therefore, several publications have included only the temperature dynamic in their models and ignored the other dynamics such as air supply and humidity. Turner *et al.* [115] and Geyer *et al.* [51] included the transient effect of fuel cell stack temperature rise in their models. By including only temperature dynamics, the system transient behavior can be clearly observed during the warm-up period as shown in [20]. Hauer *et al.* [57] represented the dynamics of the fuel reformer with the dynamics of its temperature rise by using a second-order transfer function with an adjustable time constant. Kim and Kim [66] simplified the system model further by using a first-order time delay electrical circuit to represent the fuel reformer and the fuel cell stack voltage. The model was connected to a step-up chopper. A fuzzy controller was designed to improve system performance. A few publications [56, 88, 92, 97] included the dynamics of the air supply system, that is, considered the dynamics of the air compressor and the manifold filling and their consequences to the fuel cell system.

From the literature review above, it is obvious that a comprehensive control-oriented model is needed. The field is fast evolving and there is a lot of excitement but also a lot of commercial or confidentiality considerations that do not allow state-of-the-art results to be published. The exercise of developing such a model is critical for future control development.

2

Fuel Cell System Model: Auxiliary Components

Models developed specifically for control studies have certain characteristics. Important characteristics such as dynamic (transient) effects are included and some other effects, such as spatial variation of parameters, are neglected or approximated. Furthermore, only dynamic effects that are related to the automobile control problem are integrated into the models. The relevant time constants for an automotive propulsion-sized PEM fuel cell stack system are summarized in [56] as

- Electrochemistry $O(10^{-19}$ sec), *ignored*
- Hydrogen and air manifolds $O(10^{-1}$ sec),
- Membrane water content $O($unclear$)$,
- Flow control/supercharging devices $O(10^0$ sec),
- Vehicle inertia dynamics $O(10^1$ sec), and
- Cell and stack temperature $O(10^2$ sec),

where O denotes the order of magnitude. The fast transient phenomena of electrochemical reactions have minimal effects in automobile performance and can be ignored. The transient behaviors due to manifold filling dynamics, membrane water content, supercharging devices, and temperature may have an impact on the behavior of the vehicle and, thus, must be included in the model. The response of the humidification and membrane water content cannot be easily decoupled from the temperature and flow dynamics and, thus, the associated time constant is listed as "unclear". Interactions between processes, when appropriate, are also included. However, with relatively slow responses, the cell and stack temperature may be viewed as a separate control system which is equipped with a separate controller. The temperature can then be considered as a constant for other faster subsystems.

The system block diagram showing the subsystem blocks along with input/output signals is illustrated in Figure 2.1. The thick arrow between two component blocks (marked "flow") represents flow rate as well as the condition of the gas (*e.g.*, pressure, humidity, and temperature). In this and the next chapters, the models of several components shown in the figure are explained.

We focus on the reactant supply subsystem and thus the models of the components related to this subsystem are developed. The component models for the heat management subsystem are left for future study. Figure 2.2 illustrates the components and flows related to the reactant supply subsystem. It is assumed that the cathode and anode volumes of the multiple fuel cells are lumped as a single stack cathode and anode volumes. The anode supply and return manifold volumes are very small. Their sizes allow us to lump all these volumes to one "anode" volume. The cathode supply manifold lumps all the volumes associated with pipes and connections between the compressor and the stack cathode flow field. The length, and thus volume, of the cathode supply manifold can be large depending on the physical location of the compressor with respect to the stack. The cathode return manifold represents the lumped volume of pipes downstream from the stack cathode.

    In this chapter, the modeling of the auxiliary components is explained. The compressor dynamic model is explained in Section 2.1 followed by an explanation of the manifold filling model in Section 2.2. Static models of the air cooler and the air humidifier are explained in Sections 2.4 and 2.5. In the next chapter, the development of the fuel cell stack model, which consists of stack voltage, anode flow, cathode flow, and membrane hydration models, is presented.

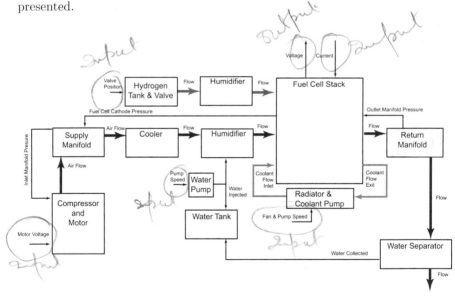

Fig. 2.1. System block diagram

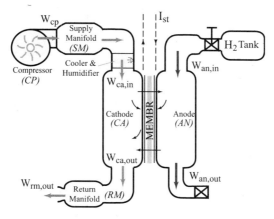

Fig. 2.2. Reactant supply subsystem model

2.1 Compressor Model

The compressor model is separated into two parts, as shown in Figure 2.3. The first part is a static compressor map which determines the air flow rate through the compressor. Thermodynamic equations are then used to calculate the exit air temperature and the required compressor power. The second part represents the compressor and motor inertia and defines the compressor speed. The speed is consequently used in the compressor map to find the air mass flow rate.

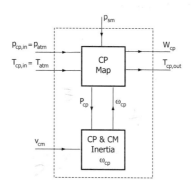

Fig. 2.3. Compressor block diagram

The only dynamic state in the model is the compressor speed ω_{cp}. The inputs to the model include inlet air pressure $p_{cp,in}$, its temperature $T_{cp,in}$, the voltage command to the compressor motor v_{cm}, and downstream pressure, which is the supply manifold pressure $p_{cp,out} = p_{sm}$. The inlet air is typically atmospheric and its pressure and temperature are assumed to be fixed at

$p_{atm} = 1$ atm and $T_{atm} = 25°C$, respectively. The motor command is one of the inputs to the fuel cell system. The downstream pressure is determined by the supply manifold model.

The compressor air mass flow rate W_{cp} (kg/sec) is determined, through a compressor flow map, from the pressure ratio across the compressor and the speed of the compressor. However, supplying the compressor flow map in the form of a look-up table is not well suited for dynamic system simulations [80]. Standard interpolation routines are not continuously differentiable and extrapolation is unreliable. Therefore, a nonlinear curve fitting method is used to model the compressor characteristics. The Jensen & Kristensen method, described in [80], is used in our model.

To reflect variations in the inlet condition of the compressor, which are the inlet flow pressure and temperature, the "corrected" values of mass flow rate and compressor speed are used in the compressor map. The corrected values [29] are the corrected compressor speed (rpm) $N_{cr} = N_{cp}/\sqrt{\theta}$, and the corrected mass flow $W_{cr} = W_{cp}\sqrt{\theta}/\delta$, where corrected temperature $\theta = T_{cp,in}/288$ K and corrected pressure $\delta = p_{cp,in}/1$ atm. Using the Jensen & Kristensen method, the dimensionless head parameter Ψ is first defined:

$$\Psi = C_p T_{cp,in} \left[\left(\frac{p_{cp,out}}{p_{cp,in}} \right)^{\frac{\gamma-1}{\gamma}} - 1 \right] / \left(\frac{U_c^2}{2} \right) \tag{2.1}$$

where the inlet air temperature $T_{cp,in}$ is in Kelvin and U_c is the compressor blade tip speed (m/s),

$$U_c = \frac{\pi}{60} d_c N_{cr} \tag{2.2}$$

d_c is the compressor diameter (m), and γ is the ratio of the specific heats of the gas at constant pressure C_p/C_v, which is equal to 1.4 in the case of air. The normalized compressor flow rate Φ is defined by

$$\Phi = \frac{W_{cr}}{\rho_a \frac{\pi}{4} d_c^2 U_c} \tag{2.3}$$

where ρ_a is the air density (kg/m^3). The normalized compressor flow rate Φ is then correlated with the head parameter Ψ by the equation

$$\Phi = \Phi_{max} \left[1 - \exp \left(\beta \left(\frac{\Psi}{\Psi_{max}} - 1 \right) \right) \right] \tag{2.4}$$

where Φ_{max}, β, and Ψ_{max} are polynomial functions of the Mach number M,

$$\begin{aligned} \Phi_{max} &= a_4 M^4 + a_3 M^3 + a_2 M^2 + a_1 M + a_0 \\ \beta &= b_2 M^2 + b_1 M + b_0 \\ \Psi_{max} &= c_5 M^5 + c_4 M^4 + c_3 M^3 + c_2 M^2 + c_1 M + c_0 \end{aligned} \tag{2.5}$$

The inlet Mach number M is defined by

Table 2.1. Compressor map parameters

Parameter	Value	Units
R_a	2.869×10^2	J/(kg·K)
ρ_a	1.23	kg/m^3
d_c	0.2286	m

$$M = \frac{U_c}{\sqrt{\gamma R_a T_{cp,in}}} \tag{2.6}$$

where R_a is the air gas constant. In Equation (2.5), a_i, b_i, and c_i are regression coefficients obtained by curve fitting of the compressor data. The air mass flow in kg/sec is then calculated using Equation (2.3):

$$W_{cr} = \Phi \rho_a \frac{\pi}{4} d_c^2 U_c \tag{2.7}$$

The parameters used in the model are given in Table 2.1. The compressor model used here is for an Allied Signal compressor. The data were obtained by digitizing the compressor map given in [29]. The regression coefficients obtained by curve fitting are given in Table 2.2. Figure 2.4 shows that the curve fitting scheme represents the compressor data very well.

Table 2.2. Compressor map regression coefficients

Parameter	Value
a_4	-3.69906×10^{-5}
a_3	2.70399×10^{-4}
a_2	-5.36235×10^{-4}
a_1	-4.63685×10^{-5}
a_0	2.21195×10^{-3}
b_2	1.76567
b_1	-1.34837
b_0	2.44419
c_5	-9.78755×10^{-3}
c_4	0.10581
c_3	-0.42937
c_2	0.80121
c_1	-0.68344
c_0	0.43331

A look-up table of the compressor efficiency η_{cp} is used to find the efficiency of the compressor from the mass flow rate and pressure ratio across the compressor. The maximum efficiency of the compressor is 80%. The temperature of the air leaving the compressor is calculated from the equation

$$T_{cp,out} = T_{cp,in} + \frac{T_{cp,in}}{\eta_{cp}} \left[\left(\frac{p_{cp,out}}{p_{cp,in}} \right)^{\frac{\gamma-1}{\gamma}} - 1 \right]$$

$$\text{\reflectbox{ρ}} = T_{atm} + \frac{T_{atm}}{\eta_{cp}}\left[\left(\frac{p_{sm}}{p_{atm}}\right)^{\frac{\gamma-1}{\gamma}} - 1\right] \tag{2.8}$$

The torque required to drive the compressor is calculated using thermodynamic equation:

$$\tau_{cp} = \frac{C_p}{\omega_{cp}}\frac{T_{atm}}{\eta_{cp}}\left[\left(\frac{p_{sm}}{p_{atm}}\right)^{\frac{\gamma-1}{\gamma}} - 1\right] W_{cp} \tag{2.9}$$

where τ_{cp} is the torque needed to drive the compressor in N-m;
C_p is the specific heat capacity of air $= 1004 \quad \text{J}\cdot\text{kg}^{-1}\cdot\text{K}^{-1}$;
γ is the ratio of the specific heats of air $= 1.4$.

Derivations of Equations (2.8) and (2.9) are standard and can be found in the thermodynamics or turbine literature [21, 53].

A lumped rotational parameter model with inertia is used to represent the dynamic behavior of the compressor speed:

$$J_{cp}\frac{d\omega_{cp}}{dt} = (\tau_{cm} - \tau_{cp}) \tag{2.10}$$

where J_{cp} is the combined inertia of the compressor and the motor (kg·m²);
ω_{cp} is the compressor speed (rad/sec);
τ_{cm} is the compressor motor torque input (N-m);
τ_{cp} is the torque required to drive the compressor (N-m).

The compressor motor torque is calculated using a static motor equation:

$$\tau_{cm} = \eta_{cm}\frac{k_t}{R_{cm}}(v_{cm} - k_v\omega_{cp}) \tag{2.11}$$

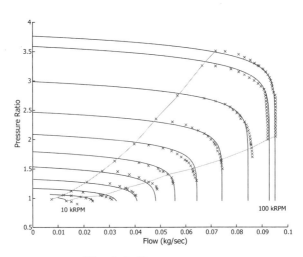

Fig. 2.4. Compressor map

where k_t, R_{cm}, and k_v are motor constants and η_{cm} is the motor mechanical efficiency. The values are given in Table 2.3.

Table 2.3. Compressor motor parameters

Parameter	Value
k_v	0.0153 V/(rad/sec)
k_t	0.0153 N-m/Amp
R_{cm}	0.82 Ω
η_{cm}	98%

2.2 Lumped Model of the Manifold Dynamics

The manifold model represents the lumped volume associated with pipes and connections between each device. The supply manifold volume includes the volume of the pipes between the compressor and the fuel cell stack including the volume of the cooler and the humidifier (Figure 1.5). The return manifold represents the pipeline at the fuel cell stack exhaust.

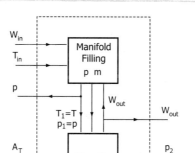

Fig. 2.5. Manifold block diagram

A block diagram of the manifold model is shown in Figure 2.5. The mass conservation principle is used to develop the manifold model. For any manifold,

$$\frac{dm}{dt} = W_{in} - W_{out} \qquad (2.12)$$

where m is the mass of the gas accumulated in the manifold volume and W_{in} and W_{out} are mass flow rates into and out of the manifold. If we assume that the air temperature is constant in the manifold T and equal to the inlet flow temperature $T = T_{in}$, the manifold filling dynamics follow an isothermic relation:

$$\frac{dp}{dt} = \frac{R_a T}{V}(W_{in} - W_{out}) \tag{2.13}$$

where R_a is the gas constant of air and V is the manifold volume. If the air temperature is expected to change in the manifold, the pressure dynamic equation, which is derived from the energy conservation, the ideal gas law, and the air thermodynamic properties,

$$\frac{dp}{dt} = \frac{\gamma R_a}{V}(W_{in}T_{in} - W_{out}T) \tag{2.14}$$

is used in addition to the mass balance equation (2.12). The air temperature T in (2.14) is calculated from the air mass m in (2.12) and air pressure p in (2.14) using the ideal gas law. In summary, if the temperature of the air in the manifold is assumed constant, Equation (2.13) is used to model the manifold dynamics. If the temperature of the air is expected to change, Equations (2.12) and (2.14) are used.

The nozzle flow equation, derived in [58], is used to calculate the outlet flow of the manifold. The flow rate passing through a nozzle is a function of the upstream pressure p_1, the upstream temperature T_1, and the downstream pressure p_2, of the nozzle. The flow characteristic is divided into two regions by the critical pressure ratio:

$$\left(\frac{p_2}{p_1}\right)_{crit} = \left(\frac{2}{\gamma+1}\right)^{\frac{\gamma}{\gamma-1}} \tag{2.15}$$

where γ is the ratio of the specific heat capacities of the gas C_p/C_v. In the case of air $\gamma = 1.4$ and the critical pressure ratio is equal to 0.528. For sub-critical flow where the pressure drop is less than the critical pressure ratio

$$\frac{p_2}{p_1} > \left(\frac{2}{\gamma+1}\right)^{\frac{\gamma}{\gamma-1}}$$

the mass flow rate is calculated from

$$W = \frac{C_D A_T p_1}{\sqrt{\bar{R}T_1}}\left(\frac{p_2}{p_1}\right)^{\frac{1}{\gamma}}\left\{\frac{2\gamma}{\gamma-1}\left[1-\left(\frac{p_2}{p_1}\right)^{\frac{\gamma-1}{\gamma}}\right]\right\}^{\frac{1}{2}} \quad \text{for } \frac{p_2}{p_1} > \left(\frac{2}{\gamma+1}\right)^{\frac{\gamma}{\gamma-1}} \tag{2.16}$$

For critical flow (or choked flow), the mass flow rate is given by

$$W_{choked} = \frac{C_D A_T p_1}{\sqrt{\bar{R}T_1}}\gamma^{\frac{1}{2}}\left(\frac{2}{\gamma+1}\right)^{\frac{\gamma+1}{2(\gamma-1)}} \quad \text{for } \frac{p_2}{p_1} \le \left(\frac{2}{\gamma+1}\right)^{\frac{\gamma}{\gamma-1}} \tag{2.17}$$

Parameter C_D is the discharge coefficient of the nozzle, A_T is the opening area of the nozzle (m^2), and $\bar{R}$ is the universal gas constant. The plot of

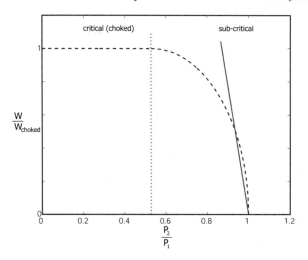

Fig. 2.6. Dashed line: relative mass flow rate as a function of nozzle pressure ratio (2.16)-(2.17); solid line: linearized mass flow rate at low pressure difference (2.18)

W/W_{choked} is shown as a dashed line in Figure 2.6. If the pressure difference between the manifold and the downstream volume is small and always falls into the subcritical flow region, the flow rate can be calculated by a linearized form of the subcritical nozzle flow equation (2.16),

$$W = k(p_1 - p_2) \qquad (2.18)$$

where k is the nozzle constant. The plot of the linearized equation (2.18) for various manifold pressures is shown in Figure 2.7 as a solid line, compared to the plot of Equation (2.16) shown as a dashed line.

2.2.1 Supply Manifold

For the supply manifold, the inlet mass flow is the compressor flow W_{cp} and the outlet mass flow is $W_{sm,out}$. Because the pressure difference between the supply manifold and the cathode is relatively small,

$$W_{sm,out} = k_{sm,out}(p_{sm} - p_{ca}) \qquad (2.19)$$

where $k_{sm,out}$ is the supply manifold outlet flow constant. Because the temperature of the air leaving the compressor is high, it is expected that the air temperature changes inside the supply manifold. Thus, Equations (2.12) and (2.14) are used to model the supply manifold

$$\frac{dm_{sm}}{dt} = W_{cp} - W_{sm,out} \qquad (2.20)$$

$$\frac{dp_{sm}}{dt} = \frac{\gamma R_a}{V_{sm}} (W_{cp}T_{cp,out} - W_{sm,out}T_{sm}) \qquad (2.21)$$

nozzle is assumed very small.

Put in m file

where V_{sm} is the supply manifold volume and T_{sm} is the supply manifold air temperature, which is calculated from m_{sm} and p_{sm} using the ideal gas law. A block diagram of the supply manifold is shown in Figure 2.8.

$Tsm =$
$\frac{psmV}{Rmsm}$

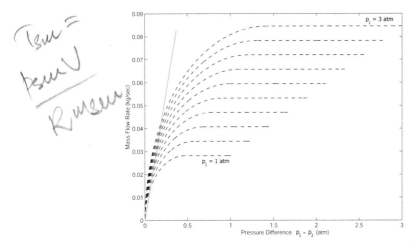

Fig. 2.7. Comparison of nozzle flow rate from nonlinear and linear nozzle equations

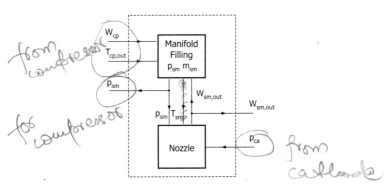

from compressor

for compressor

from cathode

Fig. 2.8. Supply manifold block diagram

2.2.2 Return Manifold

The temperature of the air leaving the stack is relatively low. Therefore, the changes of air temperature in the return manifold are negligible, and the return manifold pressure is modeled by

= leaveout

$$\frac{dp_{rm}}{dt} = \frac{R_a T_{rm}}{V_{rm}}(W_{ca,out} - W_{rm,out}) \tag{2.22}$$

where V_{rm} is the return manifold volume and T_{rm} is the temperature of the gas in the return manifold. The flow entering the return manifold $W_{ca,out}$ is calculated in Equation (3.47), which is in the same form as Equation (2.19). The outlet mass flow of the return manifold is governed by nozzle (throttle) equations (2.16) to (2.17). The outlet mass flow is a function of the manifold pressure p_{rm} and the pressure downstream from the manifold, which is assumed to be fixed at p_{atm}. Because the pressure drop between the return manifold and the atmospheric is relatively large, the equations of the return manifold exit flow are

$$W_{rm,out} = \frac{C_{D,rm} A_{T,rm} p_{rm}}{\sqrt{RT_{rm}}} \left(\frac{p_{atm}}{p_{rm}}\right)^{\frac{1}{\gamma}} \left\{\frac{2\gamma}{\gamma-1}\left[1 - \left(\frac{p_{atm}}{p_{rm}}\right)^{\frac{\gamma-1}{\gamma}}\right]\right\}^{\frac{1}{2}}$$

$$\text{for} \quad \frac{p_{atm}}{p_{rm}} > \left(\frac{2}{\gamma+1}\right)^{\frac{\gamma}{\gamma-1}} \qquad (2.23)$$

and

$$W_{rm,out} = \frac{C_{D,rm} A_{T,rm} p_{rm}}{\sqrt{RT_{rm}}} \gamma^{\frac{1}{2}} \left(\frac{2}{\gamma+1}\right)^{\frac{\gamma+1}{2(\gamma-1)}}$$

$$\text{for} \quad \frac{p_{atm}}{p_{rm}} \leq \left(\frac{2}{\gamma+1}\right)^{\frac{\gamma}{\gamma-1}} \qquad (2.24)$$

The throttle opening area $A_{T,rm}$ can be set constant or can be used as an extra control variable to regulate the return manifold pressure, and thus the cathode pressure [97]. The values of $C_{D,rm}$ and the nominal value of $A_{T,rm}$ used in the model are given in Table 4.1. A block diagram of the return manifold model is shown in Figure 2.9.

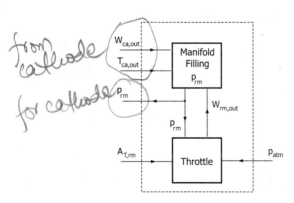

Fig. 2.9. Return manifold block diagram

The pressure calculated in the supply manifold model is used in the compressor model to determine the pressure ratio across the compressor. The return manifold pressure calculated in the return manifold model is used to determine the flow rate exiting the fuel cell cathode. The model of the cathode along with other parts of the fuel cell stack are described in the next chapter.

2.3 Review of the Thermodynamics of Gas Mixtures

In this section, we review the basic thermodynamic properties of gas mixtures that we use extensively in the model. Details can be found in [103]. We also focus on the mixture involving gases and water vapor.

Here, we consider properties of ideal gases. Specifically, each component of the mixture is independent of the presence of other components and each component can be treated as an ideal gas. Consider the mixture of gas A and gas B. From the ideal gas law, we have

$$pV = n\bar{R}T = mRT \tag{2.25}$$

where p is the gas pressure, V is the gas volume, n is the number of moles of the gas, m is the mass of the gas, $\bar{R}$ is the universal gas constant, R is the gas constant, and T is the gas temperature. The total number of moles of the mixture is equal to the sum of the number of moles of each component:

$$n = n_A + n_B \tag{2.26}$$

If we treat each component as an ideal gas, the law in (2.25) holds for each component:

$$p_A V = n_A \bar{R}T$$
$$p_B V = n_B \bar{R}T \tag{2.27}$$

where p_A and p_B are the partial pressures. By substitution of Equations (2.25) and (2.27) into Equation (2.26), we get

$$p = p_A + p_B \tag{2.28}$$

Thus, for a mixture of ideal gases, the pressure of the mixture is the sum of the partial pressures of the individual components.

Let us now consider a mixture of air and water vapor. The humidity ratio ω is defined as the ratio of the mass of water vapor m_v to the mass of dry air m_a:

$$\omega = \frac{m_v}{m_a} \tag{2.29}$$

The total mass of the mixture is $m_a + m_v$. The humidity ratio does not give a good representation of the humidity of the mixture because the maximum amount of water vapor that the air can hold (saturation) depends on the temperature and pressure of the air. The relative humidity, which represents the amount of water in the air relative to the maximum possible amount, is therefore more widely used. The relative humidity ϕ is defined as the ratio of the mole fraction of the water vapor in the mixture to the mole fraction of vapor in a saturated mixture at the same temperature and pressure. With the assumption of ideal gases, the definition reduces to the ratio of the partial pressure of the water vapor p_v in the mixture to the saturation pressure of the vapor at the temperature of the mixture p_{sat}:

$$\phi = \frac{p_v}{p_{sat}} \tag{2.30}$$

The saturation pressure p_{sat} depends on the temperature and is easily obtained from a thermodynamic table of vapor [103]. In the model, the saturation pressure is calculated from an equation of the form given in [83]. The saturation pressure data in [103] is used to obtain the coefficients in the equation:

$$\log_{10}(P_{sat}) = -1.69 \times 10^{-10}T^4 + 3.85 \times 10^{-7}T^3 - 3.39 \times 10^{-4}T^2 \\ + 0.143\,T - 20.92 \tag{2.31}$$

where the saturation pressure p_{sat} is in kPa and the temperature T is in Kelvin.

The relation between the humidity ratio and the relative humidity can be derived from the ideal gas law:

$$\omega = \frac{m_v}{m_a} = \frac{p_v V/R_v T}{p_a V/R_a T} = \frac{R_a p_v}{R_v p_a} = \frac{M_v\,p_v}{M_a\,p_a} \tag{2.32}$$

where M_v and M_a, both in kg/mol, are the molar mass of vapor and dry air, respectively. By using Equations (2.28) and (2.30), the relative humidity can be calculated from dry air pressure and the humidity ratio

$$\phi = \omega \frac{M_a\,p_a}{M_v\,p_{sat}} \tag{2.33}$$

There are some details that should be pointed out. First, relative humidity having a value of one means that the mixture is saturated or fully humidified. If there is more water content in the mixture, the extra amount of water will condense into a liquid form. Second, with the ideal gas assumption, various components in the mixture can be treated separately when performing the internal energy and enthalpy calculations.

2.4 Air Cooler (Static) Model

The temperature of the air in the supply manifold is typically high due to the high temperature of air leaving the compressor. To prevent any damage to the fuel cell membrane, the air needs to be cooled down to the stack operating temperature. In this study, we do not address heat transfer effects and thus we assume that an ideal air cooler maintains the temperature of the air entering the stack at $T_{cl} = 80°C$. It is assumed that there is no pressure drop across the cooler, $p_{cl} = p_{sm}$. Because temperature change affects gas humidity, the humidity of the gas exiting the cooler is calculated as

$$\phi_{cl} = \frac{p_{v,cl}}{p_{sat}(T_{cl})} = \frac{p_{cl}p_{v,atm}}{p_{atm}p_{sat}(T_{cl})} = \frac{p_{cl}\phi_{atm}p_{sat}(T_{atm})}{p_{atm}p_{sat}(T_{cl})} \qquad (2.34)$$

where $\phi_{atm} = 0.5$ is the nominal ambient air relative humidity and $p_{sat}(T_i)$ is the vapor saturation pressure that is a function of temperature T_i. The change in temperature does not affect the mass of the gas; thus, the mass flow rate does not change in the cooler model; that is, $W_{cl} = W_{sm,out}$.

2.5 Humidifier (Static) Model

Air flow from the cooler is humidified before entering the stack by injecting water into the air stream in the humidifier. Here, the volume of the humidifier is small and hence it can be considered as part of the supply manifold volume. A static model of the humidifier is used to calculate the change in air humidity due to the additional injected water. The temperature of the flow is assumed to be constant; thus, $T_{hm} = T_{cl}$. The injected water is assumed to be in the form of vapor or the latent heat of vaporization is assumed to be taken into account in the air cooler. Based on the condition of the flow exiting the cooler ($W_{cl} = W_{sm,out}$, p_{cl}, T_{cl}, ϕ_{cl}), the dry air mass flow rate $W_{a,cl}$, the vapor mass flow rate $W_{v,cl}$, and the dry air pressure $p_{a,cl}$, can be calculated using the thermodynamic properties discussed in Section 2.3. The vapor saturation pressure is calculated from the flow temperature using Equation (2.31). Then, the vapor pressure is determined using Equation (2.30):

$$p_{v,cl} = \phi_{cl}p_{sat}(T_{cl}) \qquad (2.35)$$

Because humid air is a mixture of dry air and vapor, dry air partial pressure is the difference between the total pressure and the vapor pressure:

$$p_{a,cl} = p_{cl} - p_{v,cl} \qquad (2.36)$$

The humidity ratio can then be calculated from

$$\omega_{cl} = \frac{M_v}{M_a}\frac{p_{v,cl}}{p_{a,cl}} \qquad (2.37)$$

where M_a is the molar mass of dry air (28.84×10^{-3} kg/mol). The mass flow rate of dry air and vapor from the cooler is

$$W_{a,cl} = \frac{1}{(1 + \omega_{cl})}W_{cl} \qquad (2.38)$$

$$W_{v,cl} = W_{cl} - W_{a,cl} \qquad (2.39)$$

The mass flow rate of dry air remains the same for the inlet and outlet of the humidifier, $W_{a,hm} = W_{a,cl}$. The vapor flow rate increases by the amount of water injected:

$$W_{v,hm} = W_{v,cl} + W_{v,inj} \tag{2.40}$$

The vapor pressure also changes and can be calculated using Equation (2.32):

$$p_{v,hm} = \omega_{cl} \frac{M_a}{M_v} p_{a,cl} = \frac{W_{v,hm}}{W_{a,cl}} \frac{M_a}{M_v} p_{a,cl} \tag{2.41}$$

The vapor pressure $p_{v,hm}$ can then be used to determine the exit flow relative humidity

$$\phi_{hm} = \frac{p_{v,hm}}{p_{sat}(T_{hm})} = \frac{p_{v,hm}}{p_{sat}(T_{cl})} \tag{2.42}$$

Because the vapor pressure increases, the total pressure also increases:

$$p_{hm} = p_{a,cl} + p_{v,hm} \tag{2.43}$$

The humidifier exit flow rate is governed by the mass continuity

$$W_{hm} = W_{a,cl} + W_{v,hm} = W_{a,cl} + W_{v,cl} + W_{v,inj} \tag{2.44}$$

The flow leaving the humidifier enters the fuel cell cathode and thus, in the next chapter, the humidifier exit flow is referred to as cathode inlet (ca, in) flow; for example, $W_{ca,in} = W_{hm}$ and $\phi_{ca,in} = \phi_{hm}$.

The models of auxiliary components in the fuel cell system are developed in this chapter. These models will interact with the fuel cell stack model. In the next chapter, the fuel cell stack model and its submodels are described.

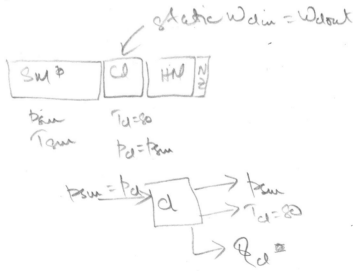

3

Fuel Cell System Model: Fuel Cell Stack

The fuel cell stack model contains four interacting submodels which are the stack voltage, the anode flow, the cathode flow, and the membrane hydration models. A block diagram of the stack model is shown in Figure 3.1. A stack thermal submodel can be added in the future when temperature changes are taken into account. The electrochemical reaction happening at the membranes is assumed to occur instantaneously. In this model, the stack temperature is assumed to be constant and uniform across the stack. Because the stack temperature dynamics has a relatively long time constant as compared to the other dynamics considered in this model, we assume that the stack temperature is constant at 80°C. The voltage model contains an equation to calculate stack voltage based on fuel cell pressure, temperature, reactant gas partial pressures, and membrane humidity. The fast dynamic effect of the electrode

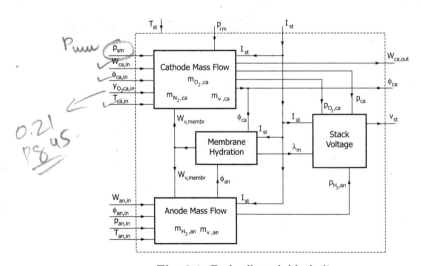

Fig. 3.1. Fuel cell stack block diagram

RC system is also explained but is not included in the model. The dynami-
cally varying pressure and relative humidity of the reactant gas flow inside the
stack flow channels are calculated in the cathode and the anode flow models.
The main flows associated with the fuel cell stack are shown in Figure 3.2
where MEA is the membrane electrode assembly that was explained in Chapter 1. The process of water transfer across the membrane is represented by
the membrane hydration model.

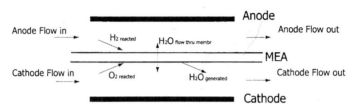

Fig. 3.2. Stack mass flow

3.1 Stack Voltage Model

In this section, the modeling of the fuel cell voltage is discussed. The voltage
is calculated as a function of stack current, cathode pressure, reactant partial
pressures, fuel cell temperature, and membrane humidity using a combination
of physical and empirical relationships. The open circuit voltage of the fuel
cell is calculated from the energy balance between chemical energy in the
reactants and electrical energy. Three main types of losses in the fuel cell are
explained. The dynamic fuel cell electrical behavior is also briefly discussed.

3.1.1 Fuel Cell Open Circuit Voltage

The fuel cell directly converts chemical energy into electrical energy. The
chemical energy released from the fuel cell can be calculated from the change
in Gibbs free energy (Δg_f) which is the difference between the Gibbs free
energy of the product and the Gibbs free energy of the reactants. The Gibbs
free energy is used to represent the available energy to do external work. For
the hydrogen/oxygen fuel cell, the basic chemical reaction is

$$H_2 + \frac{1}{2}O_2 \rightarrow H_2O \qquad (3.1)$$

and the change in the Gibbs free energy Δg_f is

$$\Delta g_f = g_f \text{ of products} - g_f \text{ of reactants} = (g_f)_{H_2O} - (g_f)_{H_2} - (g_f)_{O_2} \quad (3.2)$$

The change in Gibbs free energy varies with both temperature and pressure [69],

$$\Delta g_f = \Delta g_f^0 - \bar{R} T_{fc} \ln \left[\frac{p_{H_2} p_{O_2}^{\frac{1}{2}}}{p_{H_2O}} \right] \qquad (3.3)$$

where Δg_f^0 is the change in Gibbs free energy at standard pressure (1 bar) which varies with the temperature T_{fc} of the fuel cell, in Kelvin. The partial pressure p_{H_2}, p_{O_2}, and p_{H_2O} of the hydrogen, oxygen, and vapor, respectively, are expressed in bar. $\bar{R}$ is the universal gas constant 8.31451 J/(kg · K). The change in Gibbs free energy of the reaction in (3.1) at standard pressure Δg_f^0 is given in Table 3.1 for various reaction temperatures. The value of Δg_f^0 is negative, which means that the energy is released from the reaction.

Table 3.1. Change in Gibbs free energy of hydrogen fuel cell at various temperatures [69]

Form of Water Product	Temperature °C	Δg_f^0 (kJ/mole)
Liquid	25	-237.2
Liquid	80	-228.2
Gas	80	-226.1
Gas	100	-225.2
Gas	200	-220.4
Gas	400	-210.3
Gas	600	-199.6
Gas	800	-188.6
Gas	1000	-177.4

If the fuel cell process were "reversible," all of the Gibbs free energy would be converted to electrical energy, which is the electrical work used to move an electrical charge around a circuit. For each mole of hydrogen, two moles of electrons pass around the external circuit and the electrical work done (charge × voltage) is

$$\text{Electrical work done} = -2FE \qquad \text{Joules} \qquad (3.4)$$

where F is the Faraday Constant (= 96485 Coulombs) which represents the electric charge of one mole of electrons and E is the voltage of the fuel cell. This electrical work done would be equal to the change in Gibbs free energy if the system were considered reversible:

$$\Delta g_f = -2FE \qquad (3.5)$$

Thus, using Equation (3.3), the reversible voltage of the fuel cell can be written as

$$E = \frac{-\Delta g_f}{2F} = \frac{-\Delta g_f^0}{2F} + \frac{RT_{fc}}{2F} \ln \left[\frac{p_{H_2} p_{O_2}^{\frac{1}{2}}}{p_{H_2O}} \right] \tag{3.6}$$

In practice, the fuel cell process is not reversible; some of the chemical energy is converted to heat, and the fuel cell voltage V_{fc} is less than that in Equation (3.6). Voltage E in Equation (3.6) is called the reversible open circuit voltage or "Nernst" voltage of a hydrogen fuel cell. The term $-\Delta g_f^0/2F$ varies from standard-state (25°C and 1 atm) reference potential (1.229 V) in accordance with the temperature in the form [5],

$$-\frac{\Delta g_f^0}{2F} = 1.229 + (T_{fc} - T_0) \left(\frac{\Delta S^0}{2F} \right) \tag{3.7}$$

where T_0 is the standard-state temperature (298.15 K) and ΔS^0 is the entropy change. Because the variation in specific heat with the expected changes in temperature is minimal, the entropy change of a given reaction is approximately constant and can be set to the standard value [5]; thus,

$$-\frac{\Delta g_f^0}{2F} = 1.229 - \frac{298.15 \cdot \Delta S_0^0}{2F} + \left(\frac{\Delta S_0^0}{2F} \right) T_{fc} \tag{3.8}$$

Using thermodynamic values of the standard-state entropy change, Equation (3.8), is further expanded and yields [5]

$$E = 1.229 - 0.85 \times 10^{-3}(T_{fc} - 298.15)$$
$$+ 4.3085 \times 10^{-5} T_{fc} \left[\ln(p_{H_2}) + \frac{1}{2} \ln(p_{O_2}) \right] \quad \text{volts} \tag{3.9}$$

In Equation (3.9), T_{fc} is expressed in Kelvin, and p_{H_2} and p_{O_2} are expressed in atm.

When the fuel cell operates, the actual voltage of the cell is less than the value calculated by Equation (3.9), as shown in a typical fuel cell performance plot in Figure 3.3. The differences are a result of losses or irreversibilities. In Figure 3.3, cell voltage is the actual voltage of the fuel cell v_{cell} and the current density i is defined as cell current, which equals stack current I_{st} (A), per cell active area A_{fc} (cm^2),

$$i = \frac{I_{st}}{A_{fc}} \tag{3.10}$$

The cell current is equal to the stack current I_{st} because the stack is formed by connecting the fuel cells in series. The fuel cell losses are attributed to three categories: the activation loss, the ohmic loss, and the concentration loss. Plots of voltage drops caused by each of the losses are shown in Figure 3.4. Each of these losses is considered and modeled separately in the following sections.

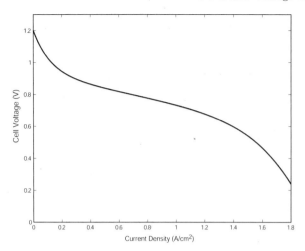

Fig. 3.3. Typical fuel cell polarization curve

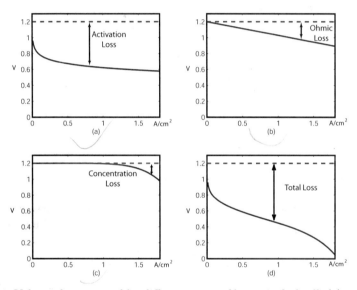

Fig. 3.4. Voltage drops caused by different types of losses in fuel cell: (a) activation losses only; (b) ohmic losses only; (c) concentration losses only; (d) total losses

3.1.2 Activation Loss

The activation loss or activation overvoltage arises from the need to move electrons and to break and form chemical bonds in the anode and cathode [73]. Part of the available energy is lost in driving the chemical reaction that transfers the electrons to and from the electrodes [69]. Activation overvoltage occurs at both fuel cell electrodes: anode and cathode. However, the reaction of hydrogen oxidation at the anode is very rapid whereas the reaction of oxygen reduction at the cathode is considerably slower [8]. Therefore, the voltage drop due to the activation loss is dominated by the cathode reaction conditions. The relation between the activation overvoltage v_{act} and the current density is described by the Tafel equation [69],

$$v_{act} = a \ln\left(\frac{i}{i_0}\right) \qquad (3.11)$$

where a is a constant and i_0, the exchange current density, is also a constant. Both constants can be determined empirically.

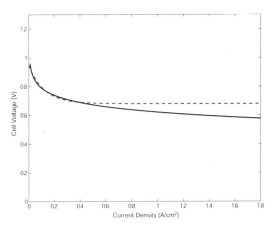

Fig. 3.5. Voltage drop caused by activation loss: solid line for (3.11); dashed line for (3.12)

The Tafel equation is, however, valid only for $i > i_0$. For low temperature PEM fuel cells, the typical value of i_0 is about 0.1 mA/cm^2 [69]. A plot of fuel cell voltage when considering only activation loss is shown as a solid line in Figure 3.5. Because Equation (3.11) is valid only for $i > i_0$, another similar function that is valid for the entire range of i is preferred in the fuel cell simulation. Therefore, the function in Equation (3.11) is approximated by

$$v_{act} = v_0 + v_a(1 - e^{-c_1 i}) \qquad (3.12)$$

where v_0 (volts) is the voltage drop at zero current density, and v_a (volts) and c_1 are constants. The activation overvoltage depends strongly on the temperature [67] and the oxygen partial pressure [5]. The values of v_0, v_a, and c_1 and their dependency on oxygen partial pressure and temperature can be determined from a nonlinear regression of experimental data using the basis function in Equation (3.12). The voltage drop calculated using Equation (3.12) is shown as a dashed line in Figure 3.5.

3.1.3 Ohmic Loss

The ohmic loss arises from the resistance of the polymer membrane to the transfer of protons and the resistance of the electrode and the collector plate to the transfer of electrons. The voltage drop that corresponds to the ohmic loss is proportional to the current density

$$v_{ohm} = i \cdot R_{ohm} \tag{3.13}$$

where R_{ohm} is the internal electrical resistance, which has units of $\Omega \cdot cm^2$. The resistance depends strongly on the membrane humidity [70] and the cell temperature [4]. Several studies in the literature [83, 104] showed that the ohmic resistance is a function of the membrane conductivity $(\Omega \cdot cm)^{-1}$, σ_m, in the form

$$R_{ohm} = \frac{t_m}{\sigma_m} \tag{3.14}$$

where t_m is the thickness of the membrane and the membrane conductivity σ_m is a function of membrane water content λ_m and fuel cell temperature. The value of λ_m between 0 and 14 corresponds to relative humidity of 0% and 100%, respectively [104]. The variation of the membrane conductivity with different membrane humidity and temperature is in the form [104]

$$\sigma_m = b_1 \exp\left(b_2 \left(\frac{1}{303} - \frac{1}{T_{fc}} \right) \right) \tag{3.15}$$

where b_1 is a function of membrane water content λ_m,

$$b_1 = (b_{11}\lambda_m - b_{12}) \tag{3.16}$$

and b_2 is a constant. Constants b_{11}, b_{12}, and b_2 are usually determined empirically. Empirical values of b_{11} and b_{12} for the Nafion 117 membrane are determined in [104].

3.1.4 Concentration Loss

Concentration loss or concentration overvoltage results from the drop in concentration of the reactants as they are consumed in the reaction. These losses are the reason for rapid voltage drop at high current density. An equation that approximates the voltage drop due to concentration losses is given by [56],

$$v_{conc} = i \left(c_2 \frac{i}{i_{max}} \right)^{c_3} \tag{3.17}$$

where c_2, c_3, and i_{max} are constants that depend on the temperature and the reactant partial pressure and can be determined empirically. The parameter i_{max} is the current density that causes precipitous voltage drop.

3.1.5 Cell Terminal Voltage

By combining all voltage drops associated with all the losses in the previous sections, the single fuel cell operating voltage can be written as

$$v_{fc} = E - v_{act} - v_{ohm} - v_{conc}$$

$$= E - \left[v_0 + v_a(1 - e^{-c_1 i})\right] - [iR_{ohm}] - \left[i\left(c_2 \frac{i}{i_{max}}\right)^{c_3}\right] \quad (3.18)$$

where the open circuit voltage E is given in Equation (3.9).

Because the fuel cell stack comprises multiple fuel cells connected in series, the stack voltage v_{st} is obtained as the sum of the individual cell voltages. Under the assumption that all cells are identical, the stack voltage can be calculated by multiplying the cell voltage v_{fc} by the number of cells n of the stack:

$$v_{st} = n v_{fc} \quad (3.19)$$

The parameters in the expression (3.18) are determined using nonlinear regression ("lsqcurvefit" of the MATLAB® optimization toolbox) on fuel cell polarization data from an automotive propulsion-sized PEM fuel cell stack [120] as shown in Figure 3.6.

The coefficients of R_{ohm} in Equation (3.14) are identified by assuming that the data are obtained from the fuel cell stack operating under a well-controlled environment, where cathode gas is fully humidified and the oxygen excess ratio is regulated at 2. Because the data plotted in Figure 3.6 are obtained from a fuel cell operated at steady-state and at designed operating conditions, it shows only the effect of different pressure and temperature on the fuel cell voltage. The effects of reactant partial pressure and membrane humidity are included by splitting their contributions to the total pressure during the regression of the activation and concentration overvoltage terms based on Equation (3.18).

Specifically, the value of b_2 is identified after setting b_1 in Equation (3.15) at the maximum membrane humidity value

$$b_1 = b_{11}(14) - b_{12} = 0.005139(14) - 0.00326 = 0.068686 \quad (3.20)$$

where $b_{11} = 0.005139$ and $b_{12} = 0.00326$ are the ones determined in [104]. The best value found for $b_2 = 350$ is different from the one given in [104]. The other parameters in Equation (3.18) are determined for every temperature and then curve fitted with the temperature.

The regression results are

$$E = 1.229 - 8.5 \times 10^{-4}(T_{fc} - 298.15)$$

$$+ 4.308 \times 10^{-5} T_{fc} \left[\ln \frac{p_{H_2}}{1.01325} + \frac{1}{2} \ln \frac{p_{O_2}}{1.01325}\right]$$

$$v_0 = 0.279 - 8.5 \times 10^{-4}(T_{fc} - 298.15)$$

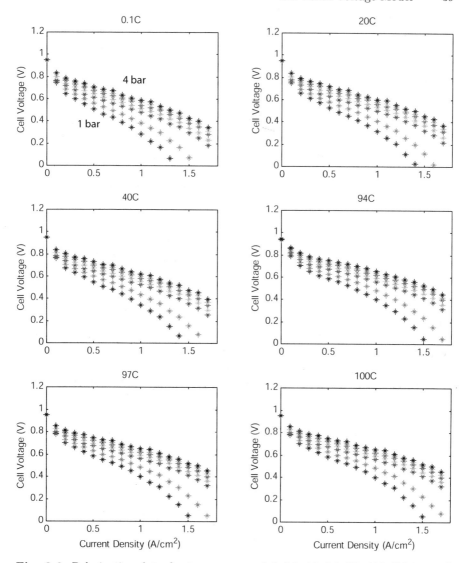

Fig. 3.6. Polarization data for temperatures 0.1, 20, 40, 94, 97, 100 Celsius and pressure 1, 1.5, 2, 2.5, 3, 3.5, 4 bar

$$+ 4.308 \times 10^{-5} T_{fc} \left[\ln \left(\frac{p_{ca} - p_{sat}}{1.01325} \right) + \frac{1}{2} \ln \left(\frac{0.1173(p_{ca} - p_{sat})}{1.01325} \right) \right]$$

$$v_a = (-1.618 \times 10^{-5} T_{fc} + 1.618 \times 10^{-2})(\frac{p_{O_2}}{0.1173} + p_{sat})^2$$

$$+ (1.8 \times 10^{-4} T_{fc} - 0.166)(\frac{p_{O_2}}{0.1173} + p_{sat}) + (-5.8 \times 10^{-4} T_{fc} + 0.5736)$$

$$c_1 = 10$$

$$t_m = 0.0125$$

$$b_1 = 0.005139\lambda_m - 0.00326$$

$$b_2 = 350$$

$$\sigma_m = b_1 \exp\left(b_2 \left(\frac{1}{303} - \frac{1}{T_{fc}} \right) \right)$$

$$R_{ohm} = \frac{t_m}{\sigma_m}$$

$$c_2 = \begin{cases} (7.16 \times 10^{-4} T_{fc} - 0.622)(\dfrac{p_{O_2}}{0.1173} + p_{sat}) \\ \quad + (-1.45 \times 10^{-3} T_{fc} + 1.68) \quad \text{for } (\dfrac{p_{O_2}}{0.1173} + p_{sat}) < 2 \text{ atm} \\ \\ (8.66 \times 10^{-5} T_{fc} - 0.068)(\dfrac{p_{O_2}}{0.1173} + p_{sat}) \\ \quad + (-1.6 \times 10^{-4} T_{fc} + 0.54) \quad \text{for } (\dfrac{p_{O_2}}{0.1173} + p_{sat}) \geq 2 \text{ atm} \end{cases}$$

$$i_{max} = 2.2$$

$$c_3 = 2 \tag{3.21}$$

where T_{fc} (K) is the temperature of the fuel cell, p_{ca} (bar) is the cathode pressure, p_{sat} (bar) is the water saturation pressure, which is a function of temperature, and p_{H_2} and p_{O_2} (bar) are the partial pressure of oxygen in the cathode and hydrogen in the anode, respectively. Examples of polarization curves created by these equations are shown in Figure 3.7. The curves of ac-

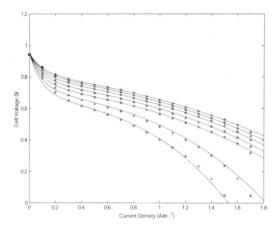

Fig. 3.7. Fuel cell polarization curve for 94°C and varying pressure from 1 to 4 bar

tivation, ohmic, and concentration overvoltage at different pressures for 80°C are shown in Figures 3.8, 3.9, and 3.10, respectively. The variation of the losses with temperature at the pressure of 2.5 bar are shown in Figures 3.11 to 3.13. The effect of membrane water content on the cell voltage is illustrated in

Figure 3.14 which shows the fuel cell polarization curve for membrane water content of 14 (100%) and 7 (50%). The model predicts significant reduction in fuel cell voltage due to change in the membrane water content. It should be noted that oversaturated (flooding) conditions will cause condensation and liquid formation inside the anode or the cathode, which leads to voltage degradation [14]. This effect is currently not captured in our model.

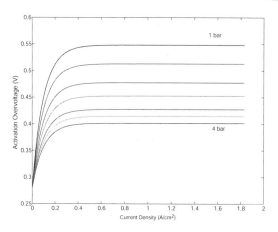

Fig. 3.8. Activation overvoltage for 80°C and pressures from 1 to 4 bar

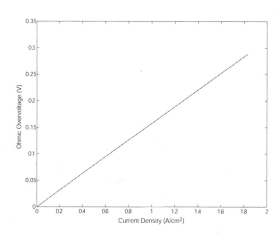

Fig. 3.9. Ohmic overvoltage for 80°C and pressures from 1 to 4 bar

The calculation of parameters in Equation (3.21) requires the knowledge of cathode pressure (representing total pressure) p_{ca}, oxygen partial pressure p_{O_2}, and fuel cell temperature T_{fc} (Figure 3.15). The pressures are calculated from the cathode model discussed in Section 3.2. The temperature can

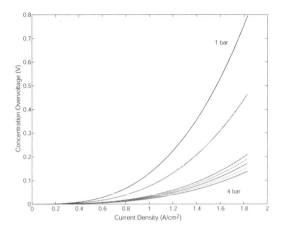

Fig. 3.10. Concentration overvoltage for 80°C and pressures from 1 to 4 bar

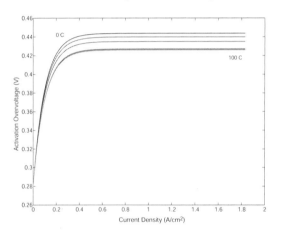

Fig. 3.11. Activation overvoltage for 2.5 bar and temperatures from 0 to 100°C

be determined based on the stack heat transfer modeling. For the current study, fixed stack temperature is assumed. The membrane conductivity that is needed in (3.14) is calculated in the membrane hydration model discussed in Section 3.4.

3.1.6 Fuel Cell Dynamic Electrical Effect

The fuel cell exhibits a fast dynamic behavior known as the "charge double layer" phenomenon [69]. Specifically, near the electrode/electrolyte interface, there is a layer of charge, called the "charge double layer," that stores electrical charge and, thus, energy. This layer behaves as an electrical capacitor. The collection of charges generates an electrical voltage that corresponds to the combination of activation overvoltage and concentration overvoltage consid-

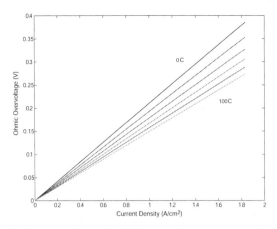

Fig. 3.12. Ohmic overvoltage for 2.5 bar and temperature from 0 to 100°C

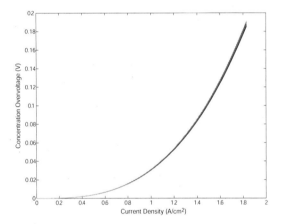

Fig. 3.13. Concentration overvoltage for 2.5 bar and temperature from 0 to 100°C

ered previously [69]. Therefore, when the current suddenly changes, it takes some time before the activation overvoltage and concentration overvoltage follow the change in the current. The ohmic voltage drop, on the other hand, responds instantaneously to a change in the current. Thus, the equivalent circuit in Figure 3.16 can be used to model the dynamic behavior of the fuel cell. Using Equations (3.12) and (3.17), we define activation resistance R_{act} and concentration resistance R_{conc} as

$$R_{act} = \frac{1}{i} \left[v_0 + v_a (1 - e^{-c_1 i}) \right] \tag{3.22a}$$

$$R_{conc} = \left(c_2 \frac{i}{i_{max}} \right)^{c_3} \tag{3.22b}$$

The dynamic fuel cell voltage behavior can be described by

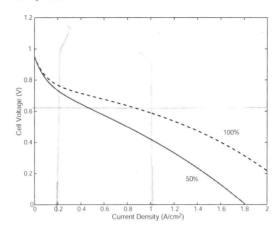

Fig. 3.14. Polarization curves for 100°C at 2.5 bar and different membrane water content

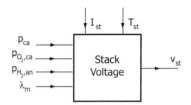

Fig. 3.15. Stack voltage model block

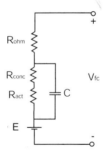

Fig. 3.16. Fuel cell equivalent circuits

$$C\frac{dv_c}{dt} + \frac{v_c - v_0}{R_{act} + R_{conc}} = i \qquad (3.23a)$$

$$v_{fc} = E - v_c - iR_{ohm} \qquad (3.23b)$$

The time constant of the fuel cell RC elements is not well established in the literature. The author in [56] reported the time constant of 10^{-19} seconds, which indicates extremely fast dynamics. This value is believed to be for a single fuel cell. The transient response of RC elements of the automobile-sized fuel cell stack can be slower. However, it is expected that the dynamics are still faster than that of the manifolds or other dynamics considered in this study. Therefore, this RC dynamic effect is not included in our model.

3.2 Cathode Flow Model

The cathode mass flow model captures the air flow behavior inside the cathode of the fuel cell stack. The model is developed using the mass conservation principle and thermodynamic and psychrometric properties of air. The thermodynamic properties of gas mixtures reviewed in Section 2.3 are used extensively in the model.

Mass continuity is used to balance the mass of three elements, namely oxygen, nitrogen, and water, inside the cathode volume, illustrated in Figure 3.17. The states of the model are oxygen mass $m_{O_2,ca}$, nitrogen mass

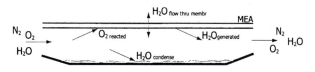

Fig. 3.17. Cathode mass flow

$m_{N_2,ca}$, and water mass $m_{w,ca}$. The subscript ca represents the fuel cell cathode. The input to the model consists of stack current I_{st}, stack temperature T_{st}, water flow rate across the membrane $W_{v,membr}$, downstream pressure, which is the return manifold pressure p_{rm}, and inlet flow properties including inlet flow temperature $T_{ca,in}$, pressure $p_{ca,in}$, mass flow rate $W_{ca,in}$, humidity $\phi_{ca,in}$, and oxygen mole fraction $y_{O_2,ca,in}$, which equals 0.21 if atmospheric air is supplied to the fuel cell. The stack temperature can be calculated using a model of stack heat transfer but it is presently assumed constant in this study. The water flow rate across the membrane is calculated by the membrane hydration model (Section 3.4), and the inlet flow properties are found in the humidifier model (Section 2.5). Figure 3.18 illustrates the calculation process in the cathode model.

Several assumptions are made. First, all gases are assumed to behave as an ideal gas. Second, the temperature of the fuel cell stack is perfectly controlled by the cooling system such that its temperature is maintained constant at 80°C and uniformly over the whole stack. Furthermore, the temperature of the flow inside the cathode flow channel is assumed to be equal to the stack temperature. Third, the variables of the flow exiting the cathode, namely, temperature $T_{ca,out}$, pressure $p_{ca,out}$, humidity $\phi_{ca,out}$, and oxygen mole fraction $y_{O_2,ca,out}$ are assumed to be the same as the variables inside the cathode flow channel, T_{ca}, p_{ca}, ϕ_{ca}, and $y_{O_2,ca}$. Therefore, following the assumptions

$$T_{ca,out} = T_{ca} = T_{st} \tag{3.24a}$$

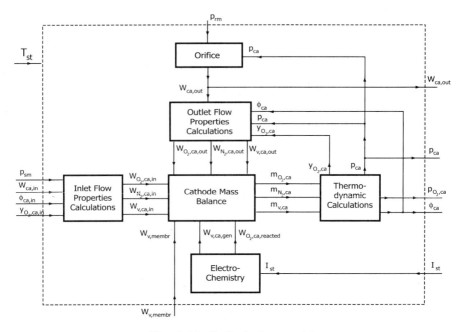

Fig. 3.18. Cathode flow model

$$p_{ca,out} = p_{ca} \tag{3.24b}$$

$$\phi_{ca,out} = \phi_{ca} \, . \tag{3.24c}$$

$$y_{O_2,ca,out} = y_{O_2,ca} \tag{3.24d}$$

Moreover, when the relative humidity of the cathode gas exceeds 100%, vapor condenses into a liquid form. This liquid water does not leave the stack and will either evaporate into the cathode gas if the gas humidity drops below 100% or it will accumulate in the cathode. Lastly, the flow channel and cathode backing layer are lumped into one volume; *i.e.*, the spatial variations are ignored.

The mass continuity is used to balance the mass of three elements – oxygen, nitrogen, and water – inside the cathode volume, as shown by

$$\frac{dm_{O_2,ca}}{dt} = W_{O_2,ca,in} - W_{O_2,ca,out} - W_{O_2,reacted} \tag{3.25}$$

$$\frac{dm_{N_2,ca}}{dt} = W_{N_2,ca,in} - W_{N_2,ca,out} \tag{3.26}$$

$$\frac{dm_{w,ca}}{dt} = W_{v,ca,in} - W_{v,ca,out} + W_{v,ca,gen} + W_{v,membr} - W_{l,ca,out} \tag{3.27}$$

where $W_{O_2,ca,in}$ is the mass flow rate of oxygen gas entering the cathode; $W_{O_2,ca,out}$ is the mass flow rate of oxygen gas leaving the cathode; $W_{O_2,reacted}$ is the rate of oxygen reacted;

$W_{N_2,ca,in}$ is the mass flow rate of nitrogen gas entering the cathode;
$W_{N_2,ca,out}$ is the mass flow rate of nitrogen gas leaving the cathode;
$W_{v,ca,in}$ is the mass flow rate of vapor entering the cathode;
$W_{v,ca,out}$ is the mass flow rate of vapor leaving the cathode;
$W_{v,ca,gen}$ is the rate of vapor generated in the fuel cell reaction;
$W_{v,membr}$ is the mass flow rate of water transfer across the membrane;
$W_{l,ca,out}$ is the rate of liquid water leaving the cathode.

All flows denoted with W terms have units of kg/sec. The inlet flow terms (subscript *in*) are calculated from the inlet flow condition (model input). The cathode outlet mass flow rate, which calculation is shown below, together with the cathode outlet gas condition is used to calculate the *out* terms. The amount of oxygen reacted and vapor produced in the reaction is calculated from the stack current using electrochemical principles. The water flow across the membrane is determined from the membrane hydration model in Section 3.4. The flow rate of liquid water leaving the cathode is zero, $W_{l,ca,out} = 0$, according to our assumptions. The calculation of the mass flow terms in the state equations (3.25) to (3.27) is explained in detail below.

The water inside the cathode volume can be in two forms, vapor and liquid, depending on the saturation state of the cathode gas. The maximum mass of vapor that the gas can hold is calculated from the vapor saturation pressure:

$$m_{v,max,ca} = \frac{p_{sat}V_{ca}}{R_v T_{st}} \tag{3.28}$$

where R_v is the gas constant of vapor. If the mass of water calculated in Equation (3.27) is more than that of the saturated state, the extra amount is assumed to condense into a liquid form instantaneously. Thus, the mass of vapor and liquid water is calculated by

$$\text{if } m_{w,ca} \leq m_{v,max,ca} \rightarrow m_{v,ca} = m_{w,ca}, \quad m_{l,ca} = 0 \tag{3.29}$$
$$\text{if } m_{w,ca} > m_{v,max,ca} \rightarrow m_{v,ca} = m_{v,max,ca}, m_{l,ca} = m_{w,ca} - m_{v,max,ca} \tag{3.30}$$

Using the masses of oxygen, nitrogen, and vapor and the stack temperature, the pressure and the relative humidity of the gas inside the cathode channel can be calculated. First, using the ideal gas law, the partial pressures of oxygen, nitrogen, and vapor inside the cathode flow channel can be calculated by

oxygen partial pressure:

$$p_{O_2,ca} = \frac{m_{O_2,ca} R_{O_2} T_{st}}{V_{ca}} \tag{3.31}$$

nitrogen partial pressure:

$$p_{N_2,ca} = \frac{m_{N_2,ca} R_{N_2} T_{st}}{V_{ca}} \tag{3.32}$$

vapor partial pressure:

$$p_{v,ca} = \frac{m_{v,ca} R_v T_{st}}{V_{ca}} \tag{3.33}$$

where R_{O_2}, R_{N_2}, and R_v are gas constants of oxygen, nitrogen, and vapor, respectively. The partial pressure of dry air is the sum of oxygen and nitrogen partial pressure.

$$p_{a,ca} = p_{O_2,ca} + p_{N_2,ca} \tag{3.34}$$

The total cathode pressure p_{ca} is the sum of air and vapor partial pressure.

$$p_{ca} = p_{a,ca} + p_{v,ca} \tag{3.35}$$

The oxygen mole fraction is determined from oxygen partial pressure and dry air partial pressure.

$$y_{O_2,ca} = \frac{p_{O_2,ca}}{p_{a,ca}} \tag{3.36}$$

The relative humidity of the cathode gas can be calculated from

$$\phi_{ca} = \frac{p_{v,ca}}{p_{sat}(T_{st})} \tag{3.37}$$

where p_{sat} is vapor saturation pressure, a function of temperature.

The inlet mass flow rate of oxygen ($W_{O_2,ca,in}$), nitrogen ($W_{N_2,ca,in}$), and vapor ($W_{v,ca,in}$) can be calculated from the inlet cathode flow condition using the thermodynamic properties discussed in Section 2.3. The saturation pressure is calculated using Equation (2.31). Then, the vapor pressure is determined using Equation (2.30):

$$p_{v,ca,in} = \phi_{ca,in} p_{sat}(T_{ca,in}) \tag{3.38}$$

Because humid air is the mixture of dry air and vapor, dry air partial pressure is therefore the difference between the total pressure and the vapor pressure.

$$p_{a,ca,in} = p_{ca,in} - p_{v,ca,in} \tag{3.39}$$

The humidity ratio is then

$$\omega_{ca,in} = \frac{M_v}{M_{a,ca,in}} \frac{p_{v,ca,in}}{p_{a,ca,in}} \tag{3.40}$$

The air molar mass M_a is calculated by

$$M_{a,ca,in} = y_{O_2,ca,in} \times M_{O_2} + (1 - y_{O_2,ca,in}) \times M_{N_2} \tag{3.41}$$

where M_{O_2} and M_{N_2} are the molar mass of oxygen and nitrogen, respectively, and $y_{O_2,ca,in}$ is 0.21 for inlet air. The mass flow rate of dry air and vapor entering the cathode is

$$W_{a,ca,in} = \frac{1}{1 + \omega_{ca,in}} W_{ca,in} \tag{3.42}$$

$$W_{v,ca,in} = W_{ca,in} - W_{a,ca,in} \tag{3.43}$$

and the oxygen and nitrogen mass flow rate can be calculated by

$$W_{O_2,ca,in} = x_{O_2,ca,in} W_{a,ca,in} \tag{3.44}$$

$$W_{N_2,ca,in} = (1 - x_{O_2,ca,in}) W_{a,ca,in} \tag{3.45}$$

where $x_{O_2,ca,in}$, defined by $x_{O_2} = m_{O_2}/m_{dryair}$, is the oxygen mass fraction, which is a function of oxygen mole fraction

$$x_{O_2,ca,in} = \frac{y_{O_2,ca,in} \times M_{O_2}}{y_{O_2,ca,in} \times M_{O_2} + (1 - y_{O_2,ca,in}) \times M_{N_2}} \tag{3.46}$$

The mass flow rate in Equations (3.43) through (3.45) are used in the state equations (3.25) to (3.27).

With the knowledge of the total flow rate at cathode exit, the mass flow rate of oxygen ($W_{O_2,ca,out}$), nitrogen ($W_{N_2,ca,out}$), and vapor ($W_{v,ca,out}$) at the exit are calculated in a similar manner as the inlet flow. The total flow rate is determined using the simplified orifice equation discussed in Section 2.2:

$$W_{ca,out} = k_{ca,out}(p_{ca} - p_{rm}) \tag{3.47}$$

where p_{ca} is the cathode total pressure, p_{rm} is the return manifold pressure (one of the model inputs), and $k_{ca,out}$ is the orifice constant. Using the mass flow rate in Equation (3.47) with conditions based on assumption (3.24), equations similar to (3.38) to (3.46) can be applied to the cathode exit flow in order to calculate $W_{O_2,ca,out}$, $W_{N_2,ca,out}$, and $W_{v,ca,out}$. The calculations are shown below in Equation (3.48). Note that, unlike the inlet flow, the oxygen mole fraction of the cathode outlet flow, which equals $y_{O_2,ca}$, is not constant because oxygen is used in the reaction. The oxygen mole fraction is calculated in Equation (3.36). The calculation of $W_{O_2,ca,out}$, $W_{N_2,ca,out}$, and $W_{v,ca,out}$ is as follows.

$$M_{a,ca} = y_{O_2,ca} \times M_{O_2} + (1 - y_{O_2,ca}) \times M_{N_2} \tag{3.48a}$$

$$\omega_{ca,out} = \frac{M_v}{M_{a,ca}} \frac{p_{v,ca}}{p_{a,ca}} \tag{3.48b}$$

$$W_{a,ca,out} = \frac{1}{1 + \omega_{ca,out}} W_{ca,out} \tag{3.48c}$$

$$W_{v,ca,out} = W_{ca,out} - W_{a,ca,out} \tag{3.48d}$$

$$x_{O_2,ca} = \frac{y_{O_2,ca} \times M_{O_2}}{y_{O_2,ca} \times M_{O_2} + (1 - y_{O_2,ca}) \times M_{N_2}} \tag{3.48e}$$

$$W_{O_2,ca,out} = x_{O_2,ca} W_{a,ca,out} \tag{3.48f}$$

$$W_{N_2,ca,out} = (1 - x_{O_2,ca}) W_{a,ca,out} \tag{3.48g}$$

Electrochemistry principles are used to calculate the rate of oxygen consumption and water production in the fuel cell reaction. The flow rate is a function of the stack current I_{st}:

$$W_{O_2,reacted} = M_{O_2} \times \frac{nI_{st}}{4F} \tag{3.49}$$

$$W_{v,ca,gen} = M_v \times \frac{nI_{st}}{2F} \tag{3.50}$$

where n is the number of cells in the stack and F is the Faraday number.

3.3 Anode Flow Model

For the system we considered, hydrogen is compressed and stored in a hydrogen tank. The high pressure storage allows us to assume that the anode inlet flow rate can be instantaneously adjusted by a valve to maintain the minimum pressure difference between the cathode and the anode. In other words, we assume that the anode channel flow resistance is small as compared to the cathode flow resistance such that maintaining the pressure difference ensures sufficient flow of hydrogen (for the fuel cell reaction). Other assumptions similar to the cathode flow model are also used. The temperature of the flow is assumed to be equal to the stack temperature. It is assumed that the anode outlet flow conditions, namely, pressure, temperature, and humidity, are the same as the conditions of the gas in the anode flow channel. Additionally, the flow channel and the backing layer of all cells are lumped into one volume.

Similar to the cathode flow model, hydrogen partial pressure and anode flow humidity are determined by balancing the mass flow of hydrogen and water in the anode. Figure 3.19 illustrates mass flow in the anode. The inputs

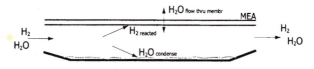

Fig. 3.19. Anode mass flow

to the model, shown in Figure 3.20, consist of anode inlet (total) mass flow $W_{an,in}$, inlet flow humidity $\phi_{an,in}$, inlet flow pressure $p_{an,in}$, inlet flow temperature $T_{an,in}$, stack current I_{st}, stack temperature T_{st}, and vapor flow rate across the membrane $W_{v,membr}$. The states are hydrogen mass $m_{H_2,an}$ and water mass $m_{w,an}$ inside the anode volume:

$$\frac{dm_{H_2,an}}{dt} = W_{H_2,an,in} - W_{H_2,an,out} - W_{H_2,reacted} \tag{3.51}$$

$$\frac{dm_{w,an}}{dt} = W_{v,an,in} - W_{v,an,out} - W_{v,membr} - W_{l,an,out} \tag{3.52}$$

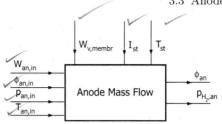

Fig. 3.20. Anode block diagram

where $W_{H_2,an,in}$ is the mass flow rate of hydrogen gas entering the anode;
$W_{H_2,an,out}$ is the mass flow rate of hydrogen gas leaving the anode;
$W_{H_2,reacted}$ is the rate of hydrogen reacted;
$W_{v,an,in}$ is the mass flow rate of vapor entering the anode;
$W_{v,an,out}$ is the mass flow rate of vapor leaving the anode;
$W_{v,membr}$ is the mass flow rate of water transfer across the membrane;
$W_{l,an,out}$ is the rate of liquid water leaving the anode.

All flows denoted with W terms have units of kg/sec. If the mass of the water calculated in Equation (3.52) is more than the maximum that the anode gas can hold, the liquid water will form inside the anode volume:

$$\text{if } m_{w,an} \leq m_{v,max,an} \rightarrow m_{v,an} = m_{w,an}, m_{l,an} = 0 \tag{3.53}$$

$$\text{if } m_{w,an} > m_{v,max,an} \rightarrow m_{v,an} = m_{v,max,an}, m_{l,an} = m_{w,an} - m_{v,max,an} \tag{3.54}$$

where the maximum vapor mass is calculated from

$$m_{v,max,an} = \frac{p_{sat}V_{an}}{R_v T_{st}} \tag{3.55}$$

The mass of the hydrogen and vapor calculated is used to determine anode pressure p_{an}, hydrogen partial pressure p_{H_2}, and the relative humidity of the gas inside the anode ϕ_{an}. The pressure is calculated using the ideal gas law.

Hydrogen partial pressure:

$$p_{H_2,an} = \frac{m_{H_2,an} R_{H_2} T_{st}}{V_{an}} \tag{3.56}$$

Vapor partial pressure:

$$p_{v,an} = \frac{m_{v,an} R_v T_{st}}{V_{an}} \tag{3.57}$$

Anode pressure:

$$p_{an} = p_{H_2,an} + p_{v,an} \tag{3.58}$$

The relative humidity of the gas inside the anode is

$$\phi_{an} = \frac{p_{v,an}}{p_{sat}(T_{st})} \tag{3.59}$$

where p_{sat} is calculated using Equation (2.31).

The inlet hydrogen mass flow $W_{H_2,an}$ and vapor mass flow $W_{v,an}$ are calculated using the anode inlet gas mass flow rate $W_{an,in}$ and humidity $\phi_{an,in}$. First, the vapor pressure is a function of the humidity:

$$p_{v,an,in} = \phi_{an,in} \cdot p_{sat}(T_{an,in}) \qquad (3.60)$$

The hydrogen partial pressure of the inlet flow is then calculated as

$$p_{H_2,an,in} = p_{an,in} - p_{v,an,in} \qquad (3.61)$$

and the anode humidity ratio is

$$\omega_{an,in} = \frac{M_v}{M_{H_2}} \frac{p_{v,an,in}}{p_{an,in}} \qquad (3.62)$$

where M_{H_2} and M_v are the molar masses of hydrogen and vapor, respectively. The mass flow rates of hydrogen and vapor entering the anode are

$$W_{H_2,an,in} = \frac{1}{1 + \omega_{an,in}} W_{an,in} \qquad (3.63)$$

$$W_{v,an,in} = W_{an,in} - W_{H_2,an,in} \qquad (3.64)$$

and are used in mass balance Equations (3.51) and (3.52). The rate of hydrogen consumed in the reaction is a function of the stack current

$$W_{H_2,reacted} = M_{H_2} \times \frac{nI_{st}}{2F} \qquad (3.65)$$

The anode exit flow rate $W_{an,out}$ represents the purge of anode gas to remove both liquid water and other gases accumulated in the anode (if reformed hydrogen is used). For the current system, it is assumed that the purge is zero. However, if the purge rate is known, the outlet hydrogen and vapor mass flow rate is calculated by the following equations.

$$\omega_{an,out} = \frac{M_v}{M_{H_2,an}} \frac{p_{v,an}}{p_{H_2,ca}} \qquad (3.66a)$$

$$W_{H_2,an,out} = \frac{1}{1 + \omega_{an,out}} W_{an,out} \qquad (3.66b)$$

$$W_{v,an,out} = W_{an,out} - W_{H_2,an,out} \qquad (3.66c)$$

It is assumed that the liquid water is stored in the anode and there is no passage available for it to leave the stack. Thus, the rate of liquid water leaving the anode $W_{l,an,out}$ is set to zero. The rate of water flow across the membrane $W_{v,membr}$ is determined in the membrane hydration model which is explained in the next section.

3.4 Membrane Hydration Model

The membrane hydration model calculates the water content in the membrane and the rate of mass flow of water across the membrane. Both water content and mass flow are assumed to be uniform over the surface area of the membrane. The membrane water content and the rate of mass flow across the membrane are functions of the stack current and the relative humidity of the flow inside the anode and the cathode flow channels (Figure 3.21). The relative humidity of the cathode and anode flow is the output of the cathode flow model and anode flow model, respectively.

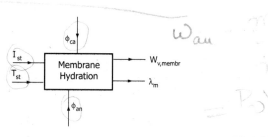

Fig. 3.21. Membrane hydration model block

The water transport across the membrane is achieved through two distinct phenomena [83, 104]:

- Water molecules are dragged across the membrane from anode to cathode by the hydrogen proton. This phenomenon is called electro-osmotic drag. The amount of water transported is represented by the electro-osmotic drag coefficient n_d, which is defined as the number of water molecules carried by each proton

$$N_{v,osmotic} = n_d \frac{i}{F} \tag{3.67}$$

where $N_{v,osmetic}$ $(mol/(sec \cdot cm^2))$ is the net water flow from anode to cathode of one cell caused by electro-osmotic drag;
i (A/cm^2) is the stack current density defined in (3.10);
F is the Faraday number.

- In a detailed spatially distributed system, there is a gradient of water concentration across the membrane that is caused by the difference in humidity in anode and cathode flows. This water concentration gradient, in turn, causes "back-diffusion" of water from cathode to anode.

$$N_{v,diff} = D_w \frac{dc_v}{dy} \tag{3.68}$$

where $N_{v,diff}$ $(mol/(sec \cdot cm^2))$ is the net water flow from cathode to anode of one cell caused by back-diffusion;

c_v (mol/cm^3) is the water concentration defined below in (3.77);

y (cm) is the distance in the direction normal to the membrane;

D_w (cm^2/sec) is the diffusion coefficient of water in the membrane.

Combining the two water transports and approximating the water concentration gradient in the membrane to be linear over the membrane thickness, the water flow across the membrane can be written as (assuming positive values in the direction from anode to cathode)

$$N_{v,membr} = n_d \frac{i}{F} - D_w \frac{(c_{v,ca} - c_{v,an})}{t_m} \tag{3.69}$$

where t_m (cm) is the thickness of the membrane. For a particular membrane, the electro-osmotic drag coefficient n_d and the diffusion coefficient D_w vary with water content in the membrane, which depends on the water content in the gas next to the membrane. Because Equation (3.69) gives the flow rate of water per unit area in (mol/(sec $\cdot$ cm^2)) in one fuel cell, the total stack mass flow rate across the membrane $W_{v,membr}$ can be calculated from

$$W_{v,membr} = N_{v,membr} \times M_v \times A_{fc} \times n \tag{3.70}$$

where M_v is the vapor molar mass, A_{fc} (cm^2) is the fuel cell active area, and n is the number of fuel cells in the stack.

The average between the water contents in the anode flow and the cathode flow can be used to represent the membrane water content. However, using the water content in the anode flow presents a more conservative approach, as discussed in [83], because the membrane water content tends to be lower on the anode side. This is because at high current density, water transport from anode to cathode by electro-osmotic drag exceeds the water back-diffusion from cathode to anode. The membrane water content, and thus the electro-osmotic and diffusion coefficients, can be calculated using the activities of the gas in the anode and the cathode:

$$a_i = \frac{y_{v,i} p_i}{p_{sat,i}} = \frac{p_{v,i}}{p_{sat,i}} \tag{3.71}$$

which, in the case of gas, is equivalent to relative humidity ϕ_i. The index i is either anode (an) or cathode (ca), $y_{v,i}$ is the mole fraction of vapor, p_i is the total flow pressure, $p_{sat,i}$ is the vapor saturation pressure, and $p_{v,i}$ is the vapor partial pressure. The water concentration in the anode flow $c_{v,an}$ and the cathode flow $c_{v,ca}$ are also functions of the activation of water in the anode flow a_{an} and in the cathode flow a_{ca}, respectively.

A summary of equations used in calculating the electro-osmotic drag coefficient, membrane water diffusion coefficient, and membrane water concentration is presented in [39]. The equations are developed based on experimental results measured for the Nafion 117 membrane in [104]. The water content in the membrane λ_i, defined as the ratio of water molecules to the number of charge sites [104], is calculated from water activities a_i (subscript i is either an-anode, ca-cathode, or m-membrane).

$$\lambda_i = \begin{cases} 0.043 + 17.81a_i - 39.85a_i^2 + 36.0a_i^3 \,, & 0 < a_i \le 1 \\ 14 + 1.4(a_i - 1) & , 1 < a_i \le 3 \end{cases} \tag{3.72}$$

where

$$a_m = \frac{a_{an} + a_{ca}}{2} \tag{3.73}$$

The membrane average water content λ_m is calculated by Equation (3.72) using the average water activity a_m between the anode and cathode water activities. The value of λ_m is used to represent the water content in the membrane. The electro-osmotic drag coefficient n_d and the water diffusion coefficient D_w are then calculated from the membrane water content[1] λ_m [39].

$$n_d = 0.0029\lambda_m^2 + 0.05\lambda_m - 3.4 \times 10^{-19} \tag{3.74}$$

and

$$D_w = D_\lambda \exp\left(2416\left(\frac{1}{303} - \frac{1}{T_{fc}}\right)\right) \tag{3.75}$$

where

$$D_\lambda = \begin{cases} 10^{-6} & , \lambda_m < 2 \\ 10^{-6}(1 + 2(\lambda_m - 2)) & , 2 \le \lambda_m \le 3 \\ 10^{-6}(3 - 1.67(\lambda_m - 3)) & , 3 < \lambda_m < 4.5 \\ 1.25 \times 10^{-6} & , \lambda_m \ge 4.5 \end{cases} \tag{3.76}$$

and T_{fc} (equals T_{st} in our model) is the temperature of the fuel cell in Kelvin. The water concentration at the membrane surfaces on anode and cathode sides, used in Equation (3.69), is a function of the membrane water content.

$$c_{v,an} = \frac{\rho_{m,dry}}{M_{m,dry}}\lambda_{an} \tag{3.77}$$

$$c_{v,ca} = \frac{\rho_{m,dry}}{M_{m,dry}}\lambda_{ca} \tag{3.78}$$

where $\rho_{m,dry}$ (kg/cm^3) is the membrane dry density and $M_{m,dry}$ (kg/mol) is the membrane dry equivalent weight.

To form the fuel cell stack model, the membrane hydration model is integrated with the stack voltage, the cathode flow, and the anode flow models developed in Sections 3.1, 3.2, and 3.3, respectively. Combining the fuel cell stack model described in this chapter with the auxiliary models described in Chapter 2 forms the dynamics of the fuel cell reactant supply system. In the next chapter, simulation results of the system model are presented and the dynamic effects of the air supply subsystem during transient operation of the fuel cell system are demonstrated.

---

[1] Membrane water content on the anode side is used in [39] because membrane dehydration is of more concern. However, we consider both membrane dehydration and membrane water flooding cases.

C = water concentration,
λ = water content in the membrane
a = water activity

Table 3.2. Thermodynamic constants used in the model

Symbol	Variable	Value
p_{atm}	atmospheric pressure	101.325 kPa
T_{atm}	atmospheric temperature	298.15 K
γ	ratio of specific heats of air	1.4
C_p	constant pressure specific heat of air	1004 J/(mol·K)
ρ_a	air density	1.23 kg/m^3
$\bar{R}$	universal gas constant	8.3145 J/(mol·K)
R_a	air gas constant	286.9 J/(kg·K)
R_{O_2}	oxygen gas constant	259.8 J/(kg·K)
R_{N_2}	nitrogen gas constant	296.8 J/(kg·K)
R_v	vapor gas constant	461.5 J/(kg·K)
R_{H_2}	hydrogen gas constant	4124.3 J/(kg·K)
M_{O_2}	oxygen molar mass	32×10^{-3} kg/mol
M_{N_2}	nitrogen molar mass	28×10^{-3} kg/mol
M_v	vapor molar mass	18.02×10^{-3} kg/mol
M_{H_2}	hydrogen molar mass	2.016×10^{-3} kg/mol
F	Faraday number	96,485 Coulombs

4

Fuel Cell System Model: Analysis and Simulation

The system model development in Chapters 2 and 3 focuses on the reactant supply systems that include the air flow control, hydrogen feed from a high-pressure tank, and the humidification of the reactant feeds. In a direct H_2 system, the PEM fuel cell becomes the main heat source. Due to the low operating temperature in PEMFCs, the dynamics of the stack temperature are considered to be relatively slow and, thus, can be viewed as a separate subsystem. As a result, the stack temperature is considered as a setpoint to the reactant systems. The control inputs are the compressor motor voltage, the hydrogen valve, and the humidifier water injection commands. In this chapter, we integrate into the model a static controller for the humidifier and a proportional controller for the hydrogen tank valve. Note here that when a fuel cell system runs based on compressed H_2 that is stored in cylinders, the air flow dynamics and the humidity management dominate the fuel cell system response. By assuming a perfect controller for the humidification, we decouple the phenomena of the air flow from the humidity. This enables us to focus on the air supply dynamics behavior and its control design. A steady-state analysis of the model presented in Section 4.2 is performed to determine the optimal air flow setpoints in terms of maximum net system power. The result corresponds with the value given in the literature as fuel cell specifications. In addition to the steady-state simulation, the dynamic model developed is also able to simulate the transient behavior of the system. The results from transient simulation are shown in Section 4.3. The transient behaviors agree with experimental data published in the literature.

The parameters used in the model are given in Table 4.1. The fuel cell stack is based on the 75 kW stacks used in the FORD P2000 fuel cell prototype vehicle [1]. The active area of the fuel cell is calculated from the peak power of the stack. The compressor model represents the Allied Signal compressor given in [29]. The membrane properties of the Nafion 117 membrane are obtained from [83]. The values of volumes are approximated from the dimensions of the P2000 fuel cell system.

Table 4.1. Parameters used in the simulations

Symbol	Variable	Value
$\rho_{m,dry}$	membrane dry density	0.002 kg/cm^3
$M_{m,dry}$	membrane dry equivalent weight	1.1 kg/mol
t_m	membrane thickness	0.01275 cm
n	number of cells in fuel cell stack	381
A_{fc}	fuel cell active area	280 cm^2
d_c	compressor diameter	0.2286 m
J_{cp}	compressor and motor inertia	5×10^{-5} kg·m^2
V_{an}	anode volume	0.005 m^3
V_{ca}	cathode volume	0.01 m^3
V_{sm}	supply manifold volume	0.02 m^3
V_{rm}	return manifold volume	0.005 m^3
$C_{D,rm}$	return manifold throttle discharge coefficient	0.0124
$A_{T,rm}$	return manifold throttle area	0.002 m^2
$k_{sm,out}$	supply manifold outlet orifice constant	0.3629×10^{-5} kg/(s·Pa)
$k_{ca,out}$	cathode outlet orifice constant	0.2177×10^{-5} kg/(s·Pa)

4.1 Humidifier and Hydrogen Flow Controls

In order to concentrate on the air supply dynamics, it is necessary to develop
controls for the anode hydrogen valve and the humidifier. A static control
of water injection in the humidifier is developed using thermodynamic calculations. The objective is to maintain the desired humidity of the air flow
entering the stack. It is assumed here that all necessary signals are available.
Proportional control is used to control the hydrogen flow, with the objective
of minimizing the pressure difference across the membrane.

4.1.1 Humidifier Control

Due to the lack of publicly available fuel cell data, the model developed in
this study has not been validated with a real fuel cell experimental system.
Several parameters are outdated, especially the parameters used in the calculation of water flow rate across the membrane (membrane hydration model),
and those that are used to represent the effect of membrane humidity to the
cell voltage (stack voltage model). As shown in Section 4.3, the model always predicts dehydration in the anode which results in considerable drops
in fuel cell voltage. Simulations under these conditions are not considered as
meaningful. Therefore, until extensive experimental data become available,
it is more appropriate to assume that the membrane is always fully humidified by other passive means, thus $\lambda_m = 14$ in (3.21). These assumptions are
justified because there is a great effort in the fuel cell industry to develop
a self-humidifying stack by redesigning stack components such as flow fields
and backing layers [106, 121]. The goal of the reactant humidity control then
becomes the regulation of the humidity of the stack inlet flow.

In the humidifier, the amount of water injected into the air flow $W_{v,inj}$ is assumed to be the exact amount that is required to maintain the desired stack inlet humidity ϕ^{des}. This amount can be calculated with the knowledge of the conditions of the humidifier inlet flow, which corresponds to the cooler exit flow (Figure 2.1). The inlet condition includes flow rate W_{cl}, temperature T_{cl}, humidity ϕ_{cl}, and pressure p_{cl}. Using Equations (2.35) to (2.39), we can calculate the dry air mass flow rate $W_{a,cl}$, the vapor mass flow rate $W_{v,cl}$, and the dry air pressure $p_{a,cl}$. Then, the flow rate of vapor injected is calculated by

$$W_{v,inj} = \frac{M_v}{M_a} \frac{\phi^{des} P_{sat}(T_{cl})}{p_{a,cl}} W_{a,cl} - W_{v,cl} \tag{4.1}$$

where M_v and M_a are the molar mass of vapor and dry air, respectively. With this assumption of perfect humidifier control, the calculation of the humidifier static model is simplified. The cathode inlet flow rate and pressure are

$$W_{ca,in} = W_{cl} + W_{v,inj} = W_{sm,out} + W_{v,inj} \tag{4.2}$$

and

$$p_{ca,in} = p_{a,cl} + \phi^{des} p_{sat}(T_{cl}) \tag{4.3}$$

respectively.

4.1.2 Hydrogen Valve Control

In the system considered in this study, hydrogen is supplied by a high-pressure tank and the flow rate is controlled by a valve. The inlet hydrogen flow is assumed to have 100% relative humidity. Due to the high-pressured storage, the hydrogen flow rate can be adjusted rapidly. With the fast actuator and fast dynamics of the anode volume, the anode hydrogen flow control can have high loop bandwidth. The goal of the hydrogen flow control is to minimize the pressure difference across the membrane, that is, the difference between anode and cathode pressures. Using simple proportional control based on the pressure difference, the pressure in the anode can quickly follow the changes in the cathode pressure. Because the valve is fast, it is assumed that the flow rate of hydrogen can be directly controlled based on the feedback of the pressure difference. However, the actual cathode and anode pressures cannot be directly measured. Thus, on the cathode side, the supply manifold pressure is used in the controller. On the anode side, because we assume that the anode supply manifold is small and its volume is lumped together with the anode volume (i.e., they have the same pressure), the anode pressure is used in the controller. The controller is in the form

$$W_{an,in} = K_1(K_2 p_{sm} - p_{an}) \tag{4.4}$$

where $K_1 = 2.1$ $(\frac{kg/s}{kPa})$ is the proportional gain and $K_2 = 0.94$ takes into account a nominal pressure drop between the supply manifold and the cathode.

The anode and cathode pressure responses for a series of current load changes in Figure 4.1 show the performance of the hydrogen flow proportional control. The anode pressure tracks the cathode pressure very well.

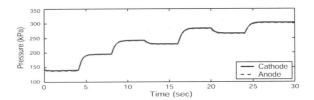

Fig. 4.1. Cathode and anode pressures in nonlinear simulation

Typically, there is also a purge valve at the end of the anode, which presents an additional control actuator. The purpose of the purge valve is to remove the liquid water accumulation in the anode to prevent flooding. Our model, however, does not incorporate water flooding effects. Therefore, the purge valve is used only to quickly reduce the anode pressure when necessary (*e.g.*, when (4.4) is negative).

4.2 Steady-state Analysis

By applying the humidifier static control and the hydrogen valve proportional control explained in the previous section, the dynamics of the fuel cell system are governed mainly by the air supply system dynamics. The air supply system has a compressor motor command as the only control actuator. Two variables considered for the control performance are the concentration of the oxygen in the cathode and the fuel cell system net power.

The net power P_{net} of the fuel cell system is the difference between the power produced by the stack P_{st} and the parasitic power required to run the auxiliary components. The majority of the parasitic power is caused by the air compressor. Therefore, it is the only parasitic loss considered in this study. For certain stack currents, the stack voltage increases with increasing air flow rate to the stack because the cathode oxygen partial pressure increases. The excess amount of air flow provided to the stack is normally indicated by the term oxygen excess ratio λ_{O_2}, defined as the ratio of oxygen supplied to oxygen used in the cathode, that is,

$$\lambda_{O_2} = \frac{W_{O_2,in}}{W_{O_2,react}} \tag{4.5}$$

High oxygen excess ratio, and thus high oxygen partial pressure, improves P_{st} and P_{net}. However, after an optimum value of λ_{O_2} is reached, further increase will cause an excessive increase in compressor power and thus deteriorate the system net power. To study the optimal value of λ_{O_2}, we plot steady-state values of λ_{O_2} and P_{net}, obtained from the simulation, for different I_{st}, as shown in Figure 4.2. For the fuel cell system modeled, the highest net power is achieved at an oxygen excess ratio between 2 and 2.4 depending on the stack current. For simplicity, it is therefore desired to control the air flow to $\lambda_{O_2} = 2$.

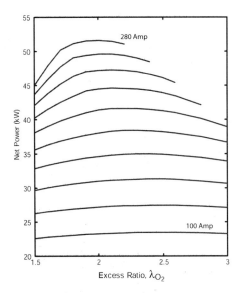

Fig. 4.2. System net power at different stack currents and oxygen excess ratios

4.3 Dynamic Simulation

A series of step changes in stack current is applied as input, as shown in Figure 4.3(a). A series of compressor motor input voltages, Figure 4.3(b), which gives different levels of steady-state oxygen excess ratios, shown in Figures 4.3(e), is also applied. This represents the simple static feedforward controller of the compressor motor based on the measurement of the current load, as shown in Figure 4.4.

During a positive current step, the oxygen excess ratio drops, as shown in Figure 4.3(e), due to the depletion of oxygen. This, in turn, causes a significant drop in the stack voltage, as shown in Figure 4.3(c). If the compressor voltage responds instantaneously during the current step (at 2, 6, 10, and 14 sec), there is still a transient effect in the stack voltage, and consequently in the stack power and the net power (Figure 4.3(c)), as a result of the transient behavior in oxygen partial pressure (Figure 4.3(f)). At $t = 18$ sec, the response

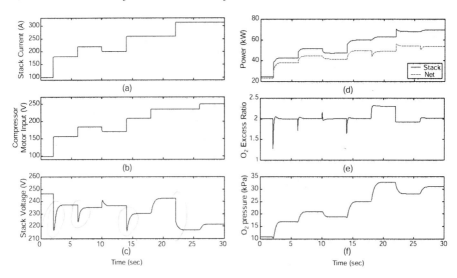

Fig. 4.3. Simulation results of the fuel cell system model for a series of input step changes

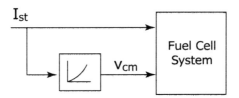

Fig. 4.4. Static feedforward using steady-state map

corresponds to a step increase in the compressor input while keeping constant stack current. An opposite case is shown at $t = 22$ sec.

The steady-state response at 16 and 20 sec shows the effect of operating at λ_{O_2} higher than the optimum value. It can be seen in Figure 4.3(c) that even though the stack power increases, the net power decreases due to the high power drawn from the compressor motor.

Figure 4.5 shows the fuel cell response on the polarization map at $80°C$. Similar results were obtained in the experiments of [70] where switching load was applied in a fuel cell. The compressor transient response is shown in Figure 4.6. Figure 4.7 shows the voltage response when considering the humidity of the membrane.

The fuel cell system model is capable of capturing the effects of transient oxygen and hydrogen partial pressures and the membrane humidity on the fuel cell voltage. Even though the model has not been validated with an actual experimental system, the model predicts transient behavior similar to that reported in the literature [70, 88]. It can be seen that the drops in fuel cell

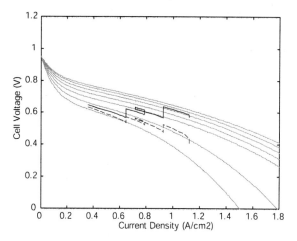

Fig. 4.5. Fuel cell response on polarization curve. Solid line assumes fully humidified membrane; dashed line represents drying membrane

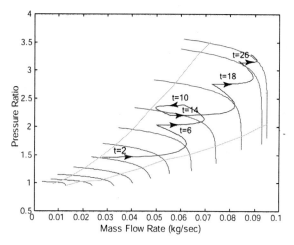

Fig. 4.6. Compressor transient response on compressor map

voltage are significant during fast load changes if the compressor motor is controlled with static feedforward based on the steady-state map. In the next chapter, a controller that gives better dynamic responses is developed using a model-based approach. Several control configurations are also considered.

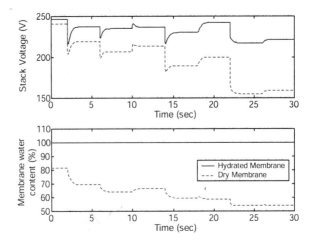

Fig. 4.7. Voltage response comparing fully humidified membrane and drying membrane

5

Air Flow Control for Fuel Cell Cathode Oxygen Reactant

There are three major control subsystem loops in the fuel cell system (FCS) that regulate the air/fuel supply, the water management, and the heat management [1]. We assume here a perfect air/fuel flow humidifier, and coolers for the incoming air and the stack. These perfect conditions are implemented in the simulation model by either fixing appropriate variables or by employing simple static controllers described in the previous chapter. Moreover, the fast proportional feedback controller on the fuel flow ensures that the anode pressure is equal to the cathode pressure following almost instantaneously any pressure variation in the cathode side. All these controllers and assumptions that are by no means trivial to implement on a real system, allow us to focus on controlling the cathode oxygen supply.

In this chapter we concentrate on the air supply subsystem of the fuel cell (FC) in order to regulate (and replenish) the oxygen depleted from the FC cathode during power generation and, in particular, current demands from the vehicle power management system. This task needs to be achieved fast and efficiently to avoid oxygen starvation and extend the life of the stack [120]. Oxygen starvation is a complicated phenomenon that occurs when the partial pressure of oxygen falls below a critical level at any possible location within the meander of the air stream in the cathode [105]. It can be observed by a rapid decrease in cell voltage that in severe cases can cause a short circuit and a hot spot on the surface of a membrane cell. Before this catastrophic event happens, the stack diagnostic system that monitors individual cell voltage removes the current from the stack or triggers system shut-down.

Although the phenomenon is spatially variant, it is largely believed that it can be avoided by regulating the excess oxygen ratio in the cathode λ_{O_2}, which is a lumped (spatially invariant) variable. This can be achieved by controlling the compressor motor to provide the air and hence the oxygen that is depleted due to the current drawn from the fuel cell. As shown in the previous chapter, there is an excess oxygen ratio that maximizes the net power from the FC system (generated FC power minus consumed compressor motor power) for each current drawn, $\lambda_{O_2}^{des} = \lambda_{O_2}^{des}(I_{st})$.

For simplification we assume for now a fixed $\lambda_{O_2}^{des} = 2$. In the future, extremum-seeking or other maximum-finding techniques can be used to search online for the optimum excess oxygen ratio levels. Note here that in a low pressure air supply system, for example, using a blower, where there are no pressure variations, regulation of λ_{O_2} corresponds to regulation of the oxygen partial pressure.

The control problem is challenging because of actuator and sensor limitations. The variables manipulated via the actuator are upstream from where the disturbance affects the performance variable (see Figure 5.1) limiting the realistic disturbance rejection capabilities of the system. Given that the exogenous input (stack current) is measured, a feedforward controller that cancels the effect of current-to-oxygen excess ratio is theoretically feasible. The design of such an ideal controller, called from now on the "cancellation" controller, is based on inverting the linearized plant model in Section 5.4. The performance and the limitations of the cancellation feedforward controller are also presented.

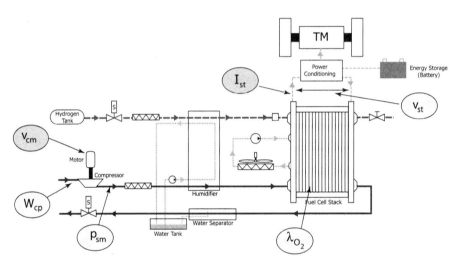

Fig. 5.1. Fuel cell system showing control inputs and outputs

In Section 5.5, a two Degrees Of Freedom (2DOF) controller is designed based on a static pre-compensator and an integral observer-based output feedback controller. The challenge here arises from the fact that not all the states, nor the performance variable λ_{O_2}, are measured. Moreover, the traditionally used measurements for λ_{O_2} regulation are upstream from the performance variable due to difficulties in sensing within a vapor-saturated flow stream. In Section 5.5.2 we demonstrate that the FCS voltage increases the system observability and thus enables a higher gain controller that improves transient λ_{O_2} regulation and robustness significantly. Currently, voltage is used in diagnostic and emergency shut-down procedures due to its fast reaction to oxygen starvation, but we clearly define its usefulness and value in a feedback design.

The FCS voltage is a natural feedback measurement for the FCS air controller and in hindsight of our results, one can view it as one of the FCS performance variables. Fast regulation of the FCS voltage to its desired value can be an indirect measure of a good level of oxygen concentration in the FC cathode. Regulating FC voltage during current demands, however, can create infeasible power setpoints and lead to instability.

Apart from being an indication of oxygen starvation, the FCS voltage is an important FC performance variable. In particular, the FCS is viewed as a power source from the DC/DC converter or other power electronics connected to it as shown in Figure 5.1. Its Current-to-Voltage transfer function defines the "power quality" of the FC as a power source [74]. The air controller designed in this section affects the closed loop Current-to-Voltage FC transfer function. We show in this chapter that the observer-based feedback controller with voltage measurement resembles a passive resistive power source; that is, for all current steps up 0.3 rad/sec, the FCS voltage behaves as $V_{st} = R_{st}I_{st}$ with a small $R_{st} = 0.05$ Ω. This result can now be used by researchers who design the power electronics for the connection of the FCS with a DC or an AC motor/generator unit for power transfer. There is also ample interest from the power generation community for the dynamics of the FCS system when connected to a grid of heterogeneous power sources [96, 74]. Many studies thus far [113] have used a static polarization curve for the Current-to-Voltage relation, which assumes a perfect or a nonexistent FC reactant flow controller.

In Section 5.7 we analyze the tradeoff between the oxygen excess ratio and the FCS system parasitic losses during transient conditions. Namely, the power utilized by the supercharger is a parasitic loss for the FC stack. We show that minimizing these parasitic losses and providing fast air flow regulation are conflicting objectives. The conflict arises from the fact that the supercharger is using part of the stack power to accelerate. One way to resolve this conflict is to augment the FC system with an auxiliary battery or an ultracapacitor that can drive the auxiliary devices or can potentially buffer the FC from transient current demands. These additional components, however, will introduce complexity and additional weight that might not be necessary [94]. To judiciously decide about the system architecture and the component sizing we analyze the tradeoff between the two objectives using linear control techniques. We then show that a compromise needs to be made between oxygen starvation and FC net power for transients faster than 0.7 rad/sec (see Figure 5.27). In other words, if net power response that is faster than 1.4 second time constant is required, our analysis suggests the use of an auxiliary power source such as a battery or capacitor. Although this answer is specific to our system, our analysis procedure is general and can be applied to other fuel cell systems.

5.1 Control Problem Formulation

As discussed in the previous chapter, the combined control design objective is to define the compressor motor input voltage v_{cm} in order to maintain $\lambda_{O_2} = 2$ and achieve the desired fuel cell system net power P_{net}^{ref}. The desired net power can be translated into the required stack current $I_{st} = f_I(P_{net}^{ref})$ assuming all other FCS variables are at the desired values. The current is then considered as an external input or disturbance to the system. The resulting control problem is defined as follows (Figure 5.2).

$$\dot{x} = f(x, u, w) \qquad \text{State Equations} \qquad (5.1)$$

$$x = \left[m_{O_2} \; m_{H_2} \; m_{N_2} \; \omega_{cp} \; p_{sm} \; m_{sm} \; m_{w,an} \; m_{w,ca} \; p_{rm} \right]^T$$

$$u = v_{cm}$$

$$w = I_{st}$$

The potential measurements include air flow rate through the compressor W_{cp}, supply manifold pressure p_{sm}, and stack voltage v_{st}.

$$y = [W_{cp} \; p_{sm} \; v_{st}]^T = h_y(x, u, w) \qquad \text{Measurements} \qquad (5.2)$$

$$z = [e_{P_{net}} \; \lambda_{O_2}]^T = h_z(x, u, w) \qquad \text{Performance Variables} \quad (5.3)$$

where $e_{P_{net}}$ is defined as the difference between the desired and the actual system net power; that is, $e_{P_{net}} = P_{net}^{ref} - P_{net}$. Figure 5.1 illustrates the physical location of all the input/output control variables.

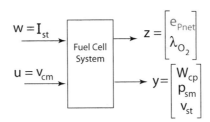

Fig. 5.2. Control problem formulation

We formulate the two control objectives $e_{P_{net}}^d = 0$ and $\lambda_{O_2}^d = 2$, but focus on the problem of using the compressor motor voltage v_{st} to regulate the oxygen excess ratio for the first sections of this chapter. Note that the two objectives are both achievable at steady-state, but their transients are considerably different, and thus cannot be achieved simultaneously by a single control actuator. The objective of achieving the desired transient system net power is ignored in the first part of the chapter, which represents the case where the power management system can rely on a secondary power source such as a battery. The tradeoff between the two performance variables, that is, $e_{P_{net}}$ and λ_{O_2}, is discussed in the last section of this chapter.

5.2 Control Configurations

The different control schemes for the fuel cell stack system are illustrated in Figure 5.3. Because the current that acts as a disturbance to λ_{O_2} can be measured, a static function that correlates the steady-state value between the control input v_{cm} and the disturbance I_{st} could be used in the feedforward path. This static feedforward is easily implemented with a look-up table (shown in Figure 5.3(a)).

The calculation of the static feedforward is based on finding the compressor voltage command v_{cm}^* that achieves the air flow which replenishes the oxygen flow that, in turn, is depleted by the reaction of hydrogen protons with oxygen molecules during a current command I_{st}. For specific ambient conditions (pressure, temperature, and humidity), the required air flow can be calculated analytically from the stack current $W_{cp}^* = f_{cp}(I_{st})$, based on electrochemical and thermodynamic principles (Section A.1). The inversion of compressor and compressor motor maps to find $v_{cm}^* = f_{cm}(I_{st})$ that gives the desired air flow W_{cp}^* is not trivial. Nonlinear simulations or testing in an experimental facility can determine the static feedforward controller from "w to u" that cancels the effect "w to z_2" at steady-state.

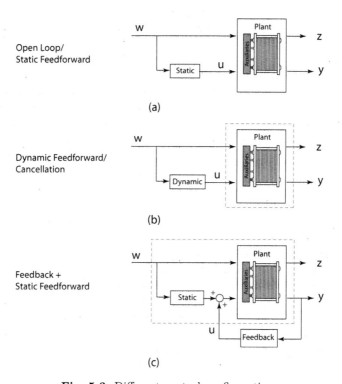

Fig. 5.3. Different control configurations

The unit step disturbance response of the system with feedforward control as compared to the system with no control is shown in Figure 5.4. The variables plotted are the deviation from nominal point $(e^d_{P_{net}}, \lambda^d_{O_2})$ of the performance variables. Because we do not have specific requirements for the $z_2 = \lambda_{O_2}$ we search for the best possible disturbance rejection that can be achieved by a controller "w to u" that cancels the disturbance from "w to z_2" at all frequencies. This cancellation controller can be implemented as a dynamic feedforward, as shown in Figure 5.3(b). The design of the dynamic feedforward controller is presented in Section 5.4. It is based on inversion of the linear plant with the input and output around the dashed area in Figure 5.3(b).

Both static and dynamic feedforward controllers suffer from sensitivity to modeling error, device aging, and variations in ambient conditions. This degrades the system robustness, that is, performance under uncertainty. To improve the system robustness, feedback control is added. Figure 5.3(c) shows the feedback control configuration. Only measurable variables y are fed back to the controller. The static feedforward is considered part of the plant. Thus, for the feedback control design, linearization is done for the inputs and outputs around the dashed box in Figure 5.3(c), which cover the static feedforward map. Note that static feedforward is used instead of dynamic feedforward. If dynamic feedforward is used, the linearization will give a higher-order plant because there are additional dynamics contributed by the dynamic feedforward.

The simplicity of the static feedforward (open loop control) with a slow proportional integral (PI) controller is very desirable, and thus establishes the basis for comparison between the performances of different controllers in the following sections. Due to the slow PI controller, the static feedforward alone defines the closed loop system behavior. Hence the response of the system with the static feedforward (open loop control) shown in Figure 5.4 is considered as the baseline controller from now on.

5.3 Linearization

The LTI system analysis in the MATLAB®/Simulink® control system toolbox is used to linearize the nonlinear system that is developed in Chapters 2 and 3. The nominal operating point is chosen where the system net power is $z^o_1 = 40$ kW and oxygen excess ratio is $z^o_2 = 2$. The inputs that correspond to this operating point are stack current at $w^o = 191$ A and compressor motor voltage at $u^o = f_{cm}(191) = 164$ V based on the static feedforward controller design discussed in the previous section. We denote also the nominal states at the equilibrium of the system for nominal inputs w^o and u^o. The linear model is given by

$$\delta \dot{x} = A \delta x + B_u \delta u + B_w \delta w$$

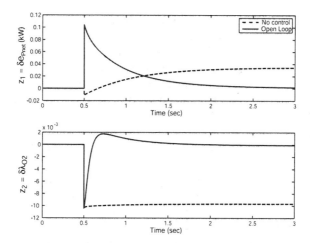

Fig. 5.4. Comparison between system with no control and system with static feedforward

$$\delta z = C_z \delta x + D_{zu} \delta u + D_{zw} \delta w \qquad (5.4)$$
$$\delta y = C_y \delta x + D_{yu} \delta u + D_{yw} \delta w$$

where $\delta(\cdot) = (\cdot) - (\cdot)^o$ represents the variation from the nominal value. The state x, measurements y, performance variables z, input u, and disturbance w are

$$x = \begin{bmatrix} m_{O_2} & m_{H_2} & m_{N_2} & \omega_{cp} & p_{sm} & m_{sm} & m_{w,an} & p_{rm} \end{bmatrix}^T \qquad (5.5)$$
$$y = [W_{cp}, p_{sm}, v_{st}]^T$$
$$z = [e_{Pnet}, \lambda_{O_2}]^T$$
$$u = v_{cm}$$
$$w = I_{st}$$

Here, the units of states and outputs are scaled such that each variable has a comparable magnitude. The units are as follows: mass in grams, pressure in bar, rotational speed in kRPM, mass flow rate in g/sec, power in kW, voltage in V, and current in A.

Note that the resulting linear model has eight states whereas the nonlinear model has nine states. The state that is removed, because it is unobservable during linearization, is the mass of water in the cathode. The reason is that the parameters of the membrane water flow we used cause excessive water flow from anode to cathode that for all nominal conditions results in fully humidified (vapor-saturated) cathode gas. Thus, for constant temperature, the vapor pressure is constant and equal to the saturated vapor pressure. Our nonlinear model does not include the effects of liquid condensation, also known as "flooding," on the FCS voltage response. As a result, the cathode water mass is not observable from the linearization point of view. On the other hand, the anode vapor pressure is observable and it is included in the linearization because variations in the FCS current affect the partial pressure of vapor in the anode that is always less than its saturated value. The change in vapor pressure affects the hydrogen partial pressure due to the fast P controller that regulates the anode pressure to be equal to the cathode pressure. The hydrogen pressure in turn affects the FCS voltage and makes the $m_{w,an}$ observable.

There are specific linearization cases. The first is the regular input/output linearization of the plant shown in Figure 5.3(b) with $(A, B_u, B_w, \ldots, D_{zw})$ as in (5.4). This is used in Section 5.4 for the design of the dynamic feedforward. The exact system matrices are given in Table A.1. The second case is linearization with static feedforward $f_{cp}(w)$ in addition to feedback control u_{fb} shown in Figure 5.3(c) $u = u_{fb} + u^* = u_{fb} + f_{cp}(w)$. The exact matrices $(A, B_u, B_w^o, \ldots, D_{zw}^o)$ are given in Table A.2 and are used in Section 5.5 where the feedback controller is designed. As our notation indicates, the matrices of the two systems are the same, except $B_w^o = \left(\frac{\partial f}{\partial w} + \frac{\partial f}{\partial u} \frac{\partial f_{cp}}{\partial w} \right) |_{x^o, u^o, w^o} = B_w + B_u \frac{\partial f_{cp}}{\partial w} |_{w^o}$ and $D_{z_1 w}^o = D_{z_1 w} + D_{z_1 u} \frac{\partial f_{cp}}{\partial w} |_{w^o}$ matrices. Note that $D_{z_2 w}$ is the same for both cases because $D_{z_2 u} = 0$.

For both linear systems, the anode flow control (proportional) is included in the linearization. The comparison between the step responses of nonlinear and linear models is shown in Figure A.1.

5.4 Dynamic Feedforward

Due to the topology of the control variable $u = v_{cm}$, and the disturbance $w = I_{st}$, with respect to the performance variable $z_2 = \lambda_{O_2}$, the disturbance rejection capabilities of the open loop system are moderate. In particular, the control signal $u = v_{cm}$ affects performance variable $z_2 = \lambda_{O_2}$ through the dynamics associated with the compressor inertia, supply manifold filling, and eventually, cathode manifold filling (see also Figure 5.1 for the physical location of the control signal). On the other hand, the disturbance $w = I_{st}$ affects the performance variable $z_2 = \lambda_{O_2}$ directly (see Figure 5.1 and Equation (4.5)). It is clear that in order to achieve good disturbance rejection the control variable u needs to be a lead filter of the measured disturbance w (see [41]). The lead filter can be based on the inversion of the open loop dynamics from "u to z_2."

Using the linear model given in Table A.1, the system can be arranged in the transfer function form

$$\Delta Z_2 = G_{z2u} \Delta U + G_{z2w} \Delta W \tag{5.6}$$

where $G_{z2u} = C_{z_2}(sI - A)^{-1} B_u$ and $G_{z2w}w = C_{z_2}(sI - A)^{-1} B_w + D_{z_2 w}$, and all variables in capital letters are in the Laplace domain. For simplicity, the Laplace variable "s" is not explicitly shown. Let a dynamic feedforward

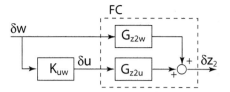

Fig. 5.5. Dynamic feedforward control

controller be $\Delta U = K_{uw}\Delta W$ as shown in Figure 5.5. The transfer function from W to Z_2 can be written as

$$T_{z_2w} = \frac{\Delta Z_2(s)}{\Delta W(s)} = (G_{z2w} + G_{z2u}K_{uw}) \tag{5.7}$$

For complete disturbance rejection $T_{z_2w} = 0$ and K_{uw}^{ideal} cancels the response of z_2 due to w:

$$K_{uw}^{ideal} = -G_{z2u}^{-1}G_{z2w} \tag{5.8}$$

The open loop plant dynamics G_{z2u} is minimum phase and thus K_{uw}^{ideal} is a stable controller. Direct modification of the current disturbance or techniques from [35] and [34] are needed in the case of a delay or nonminimum phase system dynamics. The inversion of the G_{z2u} transfer function calculated in Equation (5.8) is not proper and thus is not realizable (anticausal filter). Moreover, K_{uw}^{ideal} corresponds to a large amplitude of control input at high frequencies. To obtain a strictly proper feedforward controller, high-frequency components of K_{uw}^{ideal} are removed using a low pass filter; that is,

$$K_{uw} = -\frac{1}{(1+\dfrac{s}{\alpha_1})(1+\dfrac{s}{\alpha_2})(1+\dfrac{s}{\alpha_3})} \cdot G_{z2u}^{-1}G_{z2w} \tag{5.9}$$

The values of α_1, α_2, and α_3 used are 80, 120, and 120, respectively. Figure 5.6 shows a comparison between K_{uw}^{ideal} and the strictly proper K_{uw}.

The response of the linear system subjected to unit step in disturbance w is shown in Figure 5.7. The response of z_2 is zero except at high frequencies (*i.e.*, at the initial transient). By increasing the value of αs, the response of z_2 can be made faster at the expense of a large control action that is reflected in z_1 due to the compressor power expended.

Even though the dynamic feedforward cancels the effect of w to z_2 at a wide range of frequencies, the model-based inversion can adversely affect the disturbance rejection capability in the presence of unknown disturbance, modeling error, and parameter variation. Because there is no feedback, the sensitivity function of the system with respect to an unknown disturbance is equal to unity at all frequencies. The frequency domain modifications in [33] can be used to reduce the cancellation controller sensitivity if one can find bounds on the size of the plant uncertainties. Here, we use instead a simple PI controller that reduces the closed loop sensitivity at low frequencies

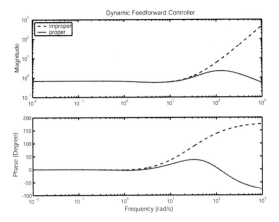

Fig. 5.6. Frequency plot of dynamic feedforward controller

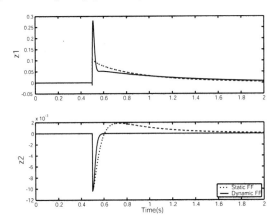

Fig. 5.7. Step response of system with dynamic feedforward in linear simulation

and ensures that W_{cp} reaches at steady-state the desired $W_{cp}^* = f_{cp}(I_{st})$; see Figure 5.8. The dFF+PI controller is given by:

$$u(t) = K_{uw}(I_{st}(t) - I_{st}^o) + k_{p,ca}(f_{cp}(I_{st}(t)) - W_{cp}(t))$$
$$+ k_{I,ca} \int_0^t (f_{cp}(I_{st}(t)) - W_{cp}(t))d\tau \quad (5.10)$$

We have observed that increasing weighting on the integrator degrades the speed of the $z_2 = \lambda_{O_2}$ response. We thus use a small integral gain $k_{I,ca}$. The fundamental reason that increasing integral gain degrades the response of performance variable $z_2 = \lambda_{O_2}$ is because the integrator is applied to the air flow measurement $y_1 = W_{cp}$ far upstream from the position where $z_2 = \lambda_{O_2}$ is defined (see additional explanation in Section 5.5.1).

Figure 5.9 shows the response of the nonlinear system with the dFF+PI subjected to a series of current steps. The dFF+PI controller has a better

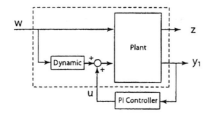

Fig. 5.8. Dynamic feedforward with PI feedback

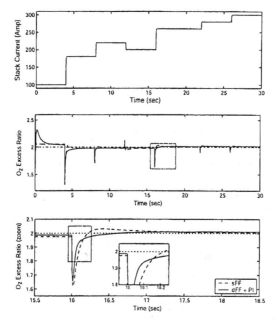

Fig. 5.9. Response of system with dynamic feedforward in nonlinear simulation

disturbance rejection capability (from $w = I_{st}$ to $z_2 = \lambda_{O_2}$) than the one achieved with the static feedforward (sFF). After the initial excursion that cannot be avoided as long as a causal controller is implemented, the dFF+PI makes λ_{O_2} recover to the 0.2% band of the nominal λ_{O_2} within 0.04 sec, whereas the sFF makes λ_{O_2} recover within 0.075 sec. This shows that the dFF+PI system is approximately two times faster than the sFF system. Note also that the overshoot in $z_2 = \lambda_{O_2}$ for the case of the sFF controller is unfavorable because redundant power is used to produce this unnecessary overshoot. Moreover, the overshoot on the O_2 excess ratio is equivalent to O_2 starvation when the system is subjected to a step-down disturbance.

The calibration and implementation of the PI controller is easy. But, the simplicity of this control configuration usually results in reduced system robustness (see Figure 5.17) as the control performance relies more on the feedforward path. In an effort to design a better (higher bandwidth) feedback controller, we explore next an observer-based feedback control design.

5.5 Feedback Control Design

Well-designed feedback controllers have advantages over feedforward control in terms of robustness in the presence of unknown disturbance and plant parameter variations. For the control problem considered here, the performance objective cannot be measured so there are inherent robustness limitations. The feedback controller is based on linear quadratic techniques decomposing the problem to a state feedback and an observer design using the separation principle. The linear model obtained from linearization with static feedforward (Table A.2) is used in designing the feedback controller.

5.5.1 State Feedback with Integral Control

Linear Quadratic Regulator (LQR) optimal control is used to design the state feedback controller. Integral control can be used together with state feedback to reduce the steady-state error of the control output. Because the performance variable λ_{O_2} cannot be measured, integral control must be applied to one of the available measurements. The most obvious choice is assigning an integrator on the compressor flow rate $y_1 = W_{cp}$ for two reasons: it is easy to measure W_{cp} and it is relatively easy to calculate the required compressor air flow rate $W_{cp}^* = f_{cp}(I_{st})$ that satisfies the desired oxygen excess ratio. This calculation (Equation (A.3)) is based on electrochemical and thermodynamic calculations for known ambient conditions. The state equation of the integrator is thus

$$\dot{q} = W_{cp}^* - W_{cp} \tag{5.11}$$

Because the control goal is to minimize the response of δz_2 without using excessive control input, the appropriate cost function is in the form

$$J = \int_0^\infty \delta z_2^T Q_z \delta z_2 + \delta u^T R \delta u \, dt \tag{5.12}$$

However, there is a disturbance feedthrough term on the performance variables δz_2:

$$\delta z_2 = C_{z2} \delta x + D_{z2w} \delta w \tag{5.13}$$

This prevents the proper formulation of the cost function in terms of the states and control signals, which is required in solving the LQR problem. In order to formulate this as an LQR problem, we first define $\delta z_2' = C_{z2} \delta x$ and use $\delta z_2'$ in the cost function as follows.

$$J = \int_0^\infty \delta z_2'^T Q_z \delta z_2' + q^T Q_I q + \delta u^T R \delta u \, dt$$

$$= \int_0^\infty \delta x^T C_{z2}^T Q_z C_{z2} \delta x + q^T Q_I q + \delta u^T R \delta u \, dt \qquad (5.14)$$

where Q_z, Q_I, and R are weighting matrices on $\delta z'$, the integrator q, and the control input u, respectively. Then, the optimal control that minimizes (5.14) is given by

$$\delta u = -K \begin{bmatrix} \delta \hat{x} & q \end{bmatrix}^T = -K_p \delta \hat{x} - K_I q \qquad (5.15)$$

where the controller gain is $K := R^{-1} \hat{B}_u^T \bar{P}$ and $\bar{P}$ denotes the solution to the Algebraic Riccati Equation (ARE):

$$\bar{P} \hat{A} + \hat{A}^T \bar{P} + \hat{Q}_x - \bar{P} \hat{B}_u R^{-1} \hat{B}_u^T \bar{P} = 0 \qquad (5.16)$$

where $\hat{A} = [A, 0 \; ; \; -C_{y1}, 0]$, $\hat{B}_u = [B_u; 0]$, $\hat{Q}_x = \mathrm{diag}(Q_x, Q_I)$, and $Q_x = C_{z2}^T Q_{z2} C_{z2}$. Due to the fact that there is a disturbance feedthrough on the performance variable (see Equation (5.13)) we add a pre-compensator u_p [42, 43] that modifies the control input:

$$\delta u = u_p - K \begin{bmatrix} \delta \hat{x}, & q \end{bmatrix}^T \qquad (5.17)$$
$$u_p = \left[C_{z_2}(A - B_u K_p)^{-1} B_u \right]^{-1} \left[D_{z_2 w} - C_{z2}(A - B_u K_p)^{-1} B_w \right] \delta w$$

Because the integrator is not used on the performance variable $\delta z_2 = \delta \lambda_{O_2}$, increasing the weighting Q_I on $q = \int (\delta W_{cp}^* - \delta W_{cp}) d\tau$ causes slow performance in terms of $\delta \lambda_{O_2}$. Figure 5.10 shows that although the high integrator gain (i.e., high weighting Q_I) brings the compressor flow rate $y_1 = W_{cp}$ to its steady-state value fast, the response of δz_2 becomes slower. This apparent tradeoff is explained below. The fast integrator regulates (to steady-state) the compressor flow that is upstream from the supply manifold and the cathode manifold. For fast recovery of λ_{O_2}, the compressor flow W_{cp} needs to exhibit overshoot. Increasing the weighting Q_z helps only the initial part of the transient, as shown in Figure 5.11. As can be seen in Figure 5.10, the best z_2 response is obtained with $Q_I = 0.001$, which gives the controller gains

$$K = \begin{bmatrix} -28.593 & -1.6 \times 10^{-13} & -60.571 & 7.572 & 579.74 & 2.55 & -3.6 \times 10^{-14} & -189.97 \end{bmatrix}$$

$$\text{and} \qquad K_I = -0.18257 \qquad (5.18)$$

This small integral gain can slowly bring the steady-state to zero. The linear responses of the system with full-state feedback controller (Equation (5.17)) and static feedforward are shown in Figure 5.12.

It can be observed that decreasing weighting on the integrator at low frequencies can improve the speed of the z_2 response because y_1 is allowed to overshoot during the transient. The frequency shaping method discussed in [7] could be applied by adding a filter to the output and augmenting the filter state in the cost function. The downside is an increase in controller complexity

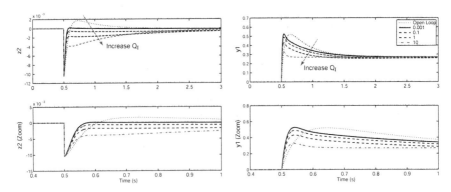

Fig. 5.10. Linear step response showing effects of increasing weight on integrator

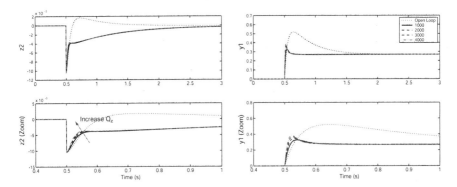

Fig. 5.11. Linear step response showing effects of increasing weight on performance variables

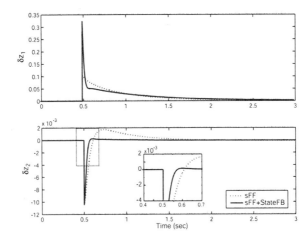

Fig. 5.12. Unit step response of system with full-state feedback in linear simulation

due to additional dynamics contributed by the output filter. The frequency shaping technique and the selection of output filter are interesting topics for future study.

Notice also that the fundamental reason that increasing integral gain degrades the response of performance variable z_2 is because the integrator is applied to the air flow measurement y_1 far upstream from the position where z_2 is defined (see Figure 5.1). Thus, moving the flow measurement to the position closer to the fuel cell stack (either flow entering or exiting) seems to be more appropriate in terms of designing integral control. However, in practice, flow measurement near the stack can cause trouble due to a large variation in the humidity, pressure, and temperature of the flow [52]. Moreover, because of this large change in the thermodynamic condition of the flow at this location, it is impossible to accurately calculate the required amount of air flow to be used as a reference value in the integral control.

In practice, to prevent stack starvation, the stack current signal is filtered by a low pass filter in order to allow enough time for the air supply system to increase air flow to the cathode. This solution, however, slows down the fuel cell power response because the power is a direct function of the current. Therefore, it is desirable to use the highest possible cutoff frequency in the low pass filter such that fast current can be drawn without starving the stack. As can be seen from Figure 5.13, to reduce the magnitude of the excess ratio, the current filter used for the controlled system can have a higher cutoff frequency, which means that the controlled system can handle faster current drawn without starving the stack and, thus, faster power is produced from the fuel cell system.

This result is easier to see in the time domain. If a current limiter were used whenever λ_{O_2} deviates, say 0.2% of the nominal value, it would have been active for 0.075 sec for the open loop system, whereas, in the closed loop system the current limiter becomes active for only 0.04 sec as seen in the zoom-in of the plot in Figure 5.12. Figure 5.14 shows that the improvements

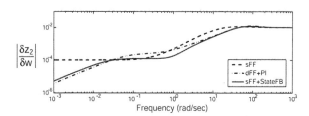

Fig. 5.13. Magnitude of frequency response of closed-loop system from w to z_2

in the closed loop performance persist in nonlinear simulations.

The plot of the input sensitivity function in Figure 5.15 shows the benefit of the feedback over the feedforward configuration. In a single-input single-output (SISO) system, the sensitivity function can be viewed as a transfer

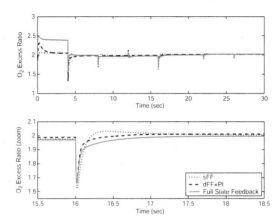

Fig. 5.14. Response of system with full-state feedback in nonlinear simulation

function from output disturbance to tracking error. Based on this interpretation, small sensitivity corresponds to good disturbance rejection. The sensitivity function is also inversely proportional to the distance between the loop gain L and the –1 point in the s-domain, $S = 1/(1 + L)$ [101]. Based on this relationship, the sensitivity function offers a measure of distance to instability. The smaller S is, the more variation in the parameters is needed to cause instability [42]. In summary, small S indicates high robustness. Figure 5.15 shows that the feedback configuration reduces system sensitivity.

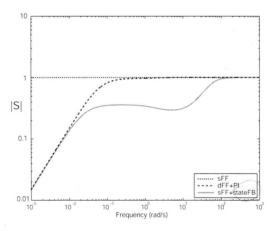

Fig. 5.15. Magnitude of input sensitivity function of system with full-state feedback

The responses shown in this section are based on the assumption that all system states are known $\delta \hat{x} = \delta x$. In practice, a state estimator (or observer) is needed to estimate the system states $\delta \hat{x}$ from available measurements y.

The design of a state observer and the effects of different measurements are presented in the following section.

5.5.2 Observer Design

The estimate of the state $\delta\hat{x}$ used in the calculation of the control input in Equation (5.17) is determined using a state observer based on the Kalman filter design. The three measurements available are compressor air flow rate $y_1 = W_{cp}$, supply manifold pressure $y_2 = p_{sm}$, and fuel cell stack voltage $y_3 = v_{st}$ (see Figure 5.1). These variables are relatively easy to measure. The compressor flow rate is typically measured and used for feedback to the compressor controller. The stack voltage is normally monitored for diagnostics and fault detection. Low voltage indicates a fault and triggers FCS shut-down and removes the current drawn from the FCS. The observer state equations are

$$\delta\dot{\hat{x}} = A\delta\hat{x} + B_u\delta u + B_w^o\delta w + L(\delta y - \delta\hat{y}) \tag{5.19}$$
$$\delta\hat{y} = C_y\delta\hat{x} + D_{yu}\delta u + D_{yw}\delta w$$

Based on the linear quadratic Gaussian method , the optimal observer gain L is

$$L := SC_y^T W_y^{-1} \tag{5.20}$$

where S is the solution to

$$0 = SA^T + AS + V_x + SC_y^T W_y^{-1} C_y S \tag{5.21}$$

The positive definite matrices W_y and V_x represent the intensities of measurement noise and process disturbance, respectively [38].

The observability analysis is summarized in Table 5.1, which shows the system eigenvalues λ_i, the corresponding eigenvectors, the rank, and the condition number of

$$\begin{bmatrix} \lambda_i I - A \\ C_y \end{bmatrix} \tag{5.22}$$

for several different cases: (1) measuring only y_1, (2) measuring y_1 and y_2, and (3) measuring all y_1, y_2, and y_3. The dynamics associated with an eigenvalue is unobservable if the corresponding matrix (5.22) loses rank (Section 2.4 of [63]). In this sense the corresponding eigenvalue can be called unobservable. A large condition number of a matrix implies that the matrix is almost rank deficient. Thus, the large condition number of the matrix (5.22) indicates a weakly observable eigenvalue λ_i.

Comparing cases (1) and (2), Table 5.1 shows that adding the y_2 measurement does not change the observability. This is because pressure and flow are related with only an integrator. The eigenvalues -219.63 and -22.404 are not observable with measurements y_1 and y_2. The eigenvectors associated with

Table 5.1. Eigenvalues, eigenvectors, and observability

Eigenvalues								
λ	-219.63	-89.485	-46.177	-22.404	-18.258	-2.915	-1.6473	-1.4038
Eigenvectors								
x1	1.06E-16	-0.17539	-0.091325	3.43E-16	0.050201	0.024367	0.86107	-0.25619
x2	0.29428	0.016479	0.012583	0.1289	0.0036888	0.016047	0.007579	-0.0074482
x3	-3.23E-16	-0.74707	-0.099392	-5.92E-16	0.13993	0.44336	-0.14727	-0.098068
x4	-1.21E-16	-0.12878	-0.45231	3.24E-15	-0.98678	0.62473	0.27811	-0.27037
x5	-9.58E-18	0.0479	0.067229	-5.98E-17	0.0046179	0.046501	0.022519	-0.022231
x6	-7.23E-17	0.61398	0.86233	-7.93E-16	0.057898	0.6389	0.3981	-0.92234
x7	0.95572	0.071474	0.11197	-0.99166	-0.016026	-0.0078755	-0.0026628	0.0024275
x8	-3.04E-17	0.099469	-0.12794	-2.05E-16	0.022705	0.043444	0.021407	-0.019503
Measuring y1								
Rank(λI-A; C)	7	8	8	7	8	8	8	8
Cond(λI-A; C)	1.29E+16	171.17	157.79	9.52E+16	461.59	1130.3	9728.4	2449.9
Measuring y1 y2								
Rank(λI-A; C)	7	8	8	7	8	8	8	8
Cond(λI-A; C)	1.32E+16	171.16	157.79	3.15E+17	461.59	1130.3	9728.4	2449.9
Measuring y1 y2 y3								
Rank(λI-A; C)	8	8	8	8	8	8	8	8
Cond(λI-A; C)	226.69	154.99	143.86	943.77	402.8	938.86	1617	1886.2

these eigenvalues suggested that the unobservable mode is almost a representation of the mass of vapor in the anode $m_{w,an}$. This agrees with the fact that the two measurements are in the air supply side and the only connection to the water in the anode is small membrane water flow. The hydrogen mass is however (more) observable through the anode flow control (which regulates anode pressure following cathode pressure). These two unobservable eigenvalues are, however, fast and thus have small effect on observer performance. On the other hand, the slow eigenvalues at −1.6473 and −1.4038 can degrade observer performance because they are weakly observable, as indicated by the large condition numbers at 9728.4 and 2449.9, respectively.

Adding the stack voltage measurement improves the state observability, as can be seen from the rank and the condition number for case 3. However, the high condition number for a slow eigenvalue (−1.4038) could degrade observer performance. Many design iterations confirm the degradation. When this eigenvalue is moved, the resulting observer gain is large, and thus produces large overshoot in observer error. From the implementation viewpoint, when combined with a controller, large observer gain can produce a compensator with undesirably high gain. To prevent high observer gain, we design a reduced-order output estimator (closed-loop observer) for the observable part and an input estimator (open-loop observer) for the weakly observable part. Below, the design process for the case of three measurements is explained.

First, the system matrices are transformed to the modal canonical form $\delta\tilde{x} = T\delta x$ [26] such that the new system matrices are

$$\tilde{A} = TAT^{-1} = \begin{bmatrix} \lambda_1 & & 0 \\ & \ddots & \\ 0 & & \lambda_8 \end{bmatrix} \qquad (5.23)$$

$$\tilde{C} = C_y T^{-1} \qquad\qquad \tilde{B} = T\,[\,B_w\ B_u\,] \qquad (5.24)$$

Note the special structure of matrix $\tilde{A}$ which has eigenvalues on the diagonal. The matrices are then partitioned into

$$\begin{bmatrix} \tilde{A}_o & 0 \\ 0 & \tilde{A}_{\bar{o}} \end{bmatrix} \qquad \begin{bmatrix} \tilde{B}_o \\ \tilde{B}_{\bar{o}} \end{bmatrix} \qquad [\,\tilde{C}_o\ \tilde{C}_{\bar{o}}\,] \qquad (5.25)$$

where $\tilde{A}_{\bar{o}} = \lambda_8 = -1.4038$. The reduced-order observer gain $\tilde{L}$ is then designed for matrices $\tilde{A}_o$, $\tilde{B}_o$, and $\tilde{C}_o$.

$$\tilde{L} := \tilde{S}\tilde{C}_o^T \tilde{W}_y^{-1} \qquad (5.26)$$

with

$$0 = \tilde{S}\tilde{A}_o^T + \tilde{A}_o\tilde{S} + \tilde{V}_x + \tilde{S}\tilde{C}_o^T \tilde{W}_y^{-1}\tilde{C}_o\tilde{S} \qquad (5.27)$$

The chosen weighting matrices are

$$\tilde{V}_x = \mathrm{diag}[\,0.01\ 10\ 10\ 0.01\ 10\ 10\ 10\ ] + \alpha\,\tilde{B}_o\,\tilde{B}_o^T \qquad (5.28)$$

$$\tilde{W}_y = 1\times 10^{-6}\,\mathrm{diag}\,[\,10\ 100\ 1\,] \qquad (5.29)$$

which correspond to the process noise and to the measurement noise, respectively, in the stochastic Kalman estimator design [7]. The $\tilde{V}_x$ is in the form used in the feedback loop recovery procedure [38]. Using this procedure, the full state feedback loop gain properties can be recovered by increasing the value of α. The value of α chosen in this design is 30. The reduced-order observer gain $\tilde{L}$ is then transformed to the original coordinate

$$L = T^{-1}\begin{bmatrix} \tilde{L} \\ 0 \end{bmatrix} \qquad (5.30)$$

Figure 5.16 (right) shows the response of observer error based on three measurements in linear simulation. The initial errors of all states are set at 1% of maximum deviation from the nominal point. It can be seen that most of the states converge within 0.4 sec. There is one slow convergence which is caused by the weakly observable eigenvalue ($\lambda_8 = -1.4038$). Figure 5.16 (left) shows the observer response when using one measurement, $y_1 = W_{cp}$. Large overshoot and slow convergence can be observed.

Figure 5.17 shows that the single measurement (i.e., $y_1 = W_{cp}$) feedback (with static feedforward (sFF)) cannot reduce the input sensitivity function as much as the multiple measurement feedback (+sFF) can. The loop transfer recovery method [38] could be used to bring the input sensitivity closer to that of full-state feedback. However, the convergence rate of the observer

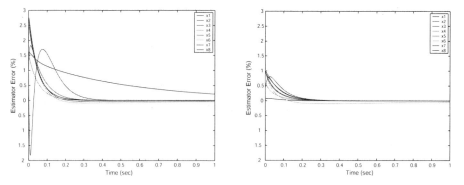

Fig. 5.16. Observer state error using only the first measurement (left) and all three measurements (right)

becomes much slower. The single measurement feedback (+sFF) has better bandwidth than the dFF+PI controller but the full potential of the model-based controller is realized when the voltage measurement $y_3 = v_{st}$ is included in the feedback. In particular, Figure 5.17 shows that the feedback with three measurements fully recovers the robustness of full-state feedback.

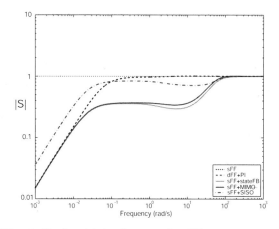

Fig. 5.17. Sensitivity function for different controllers

Simulations of the nonlinear system with different controllers are shown in Figure 5.18. Good transient response is achieved by both dynamic feedforward control (dFF+PI) and combined static feedforward and feedback with three measurements. The feedback configuration is, however, superior in terms of robustness. The analysis of the feedback controller performance and robustness indicates that the voltage measurement should be used as feedback to the controller and not only for safety monitoring.

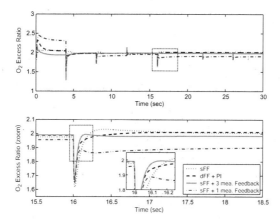

Fig. 5.18. Nonlinear simulation of system with different controllers

Different control configurations are considered in this first part of the chapter. The control design and discussion of the features and properties of each control design are presented. The advantages and disadvantages, such as simplicity and robustness, of each configuration are succinctly explained. Depending on the characteristics of the fuel cell system and the system model (for example, the source of unknown disturbance, the degree of parameter variations, and/or model accuracy) a control engineer can select the most suitable control configuration. Because of its good performance and robustness, the observer-based feedback with the FCS voltage measurement is used in the remaining sections.

5.6 Closed Loop Fuel Cell Impedance

The closed loop fuel cell system is comprised so far of (i) the air flow controller with the observer-based feedback described above, (ii) the simple PI anode pressure controller, and (iii) the perfect cathode humidification described in Chapter 4. Figure 5.19 shows a schematic of the closed loop configuration with emphasis on the air flow controller. The closed loop FC system is viewed as a voltage source from the power management system, as shown in Figure 5.20.

The voltage of the controlled FCS (cFCS) can be written as $v_{st}(t) = v_{st}^o + \mathcal{L}^{-1}\left(Z_{cFCS}(s)\Delta I_{st}(s)\right)$ where $Z_{cFCS}(s)$ is the impedance of the cFCS and $\mathcal{L}$ is the Laplace transformation. Figure 5.21 shows the $Z_{cFCS}(s)$ in a Bode magnitude and phase plot. Electrochemical impedances are sometimes also shown with Nyquist plots (see, for example, [89, 118]) and used to identify the FCS performance for different material selection. The Bode plot in Figure 5.21 indicates that the cFCS can be represented by a passive resistance $\min\left(|Z_{cFCS}|\right) = R_{cFCS}^{min} = 0.05\ \Omega$ for current commands slower than 0.1 rad/sec. A passive resistance of $\max\left(|Z_{cFCS}|\right) \approx R_{cFCS}^{max} = 0.3\ \Omega$ can also

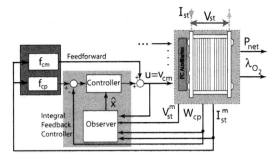

Fig. 5.19. Controlled fuel cell stack as viewed from the power management system

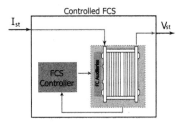

Fig. 5.20. Controlled fuel cell stack modeled as impedance

be used for current commands faster than 100 rad/sec. From the impedance phase, one can clearly see that the voltage drops for increasing current.

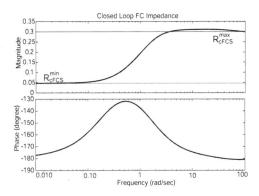

Fig. 5.21. Impedance of the controlled fuel cell stack

A plot of current–voltage trajectories against the polarization curves of the noncontrolled FCS is shown in Figure 5.22. Immediately after a step change in current the voltage drops along the fixed cathode pressure polarization curve based on the high frequency impedance ($R_{cFCS}^{max} = 0.3 \; \Omega$). After the initial transient, the controlled FCS shows a voltage that transverses to another polarization curve of higher cathode pressure. This behavior justifies the smaller cFCS resistance ($R_{cFCS}^{min} = 0.05\ \Omega$) at low frequencies. The increase in operating cathode pressure is dictated by the λ_{O_2} regulation. This phenomenon is associated with the high-pressure air supply through a high-speed compressor. A low-pressure FCS will have similar controlled and uncontrolled impedances, primarily due to the approximately constant operating pressure. Figure 5.23

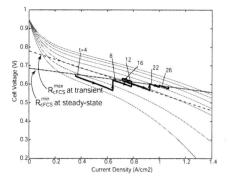

Fig. 5.22. Current–voltage trajectories that correspond in the nonlinear simulation of Figure 5.18 plotted versus the open loop FCS polarization curves

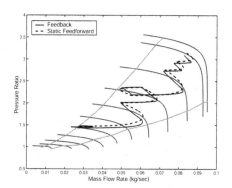

Fig. 5.23. Compressor flow/pressure trajectories during the nonlinear simulation of Figure 5.18 plotted against the compressor map

shows the compressor flow/pressure trajectories during the nonlinear simulation of Figure 5.18 plotted against the compressor map. This plot shows the actuator activity and indicates which current steps can bring the compressor close to surge or stall conditions.

5.7 Tradeoff Between Two Performance Objectives

In the case when there is no additional energy storage device such as a battery or supercapacitor, the power used to run the compressor motor needs to be taken from the fuel cell stack. A transient step change in stack current requires rapid increase in air flow to prevent depletion of cathode oxygen. This requires power drawn by the compressor motor (P_{cm}) and thus an increase in parasitic loss, which affects the system net power ($P_{net} = P_{FC} - P_{cm}$).

The control problem that we have considered so far is the single-input single-output problem of controlling the compressor command $u = v_{cm}$ to regulate the oxygen excess ratio $z_2 = \lambda_{O_2}$. During steady-state, achieving the desired value of $z_2 = \lambda_{O_2}$ ensures that the desired net power $z_1 = P_{net}$ is obtained. During transients, however, the two objectives are independent, resulting in a Single-Input Two-Output (SITO) control problem [44] shown in Figure 5.24.

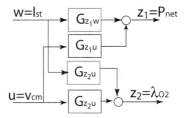

Fig. 5.24. Schematic of the coupling from I_{st} and v_{cm} to the performance variables P_{net} and λ_{O_2}

Let us consider first the effects of the exogenous input $w = I_{st}$ and the control signal $u = v_{cm}$ on the first performance variable $z_1 = P_{net}(I_{st}, v_{cm}) = P_{FC}(I_{st}, v_{cm}) - P_{cm}(v_{cm})$, or in the linear sense $\delta z_1 = G_{z_1 w} \delta w + G_{z_1 u} \delta u$. As can be seen from the step responses in Figure 5.25, I_{st} has a positive effect on the net power. On the other hand, the compressor command v_{cm} causes an initial inverse response in the net power due to a nonminimum phase zero. The last plot in Figure 5.25 shows the net power during a step change in I_{st} together with a step change in $v_{cm} = f_{cm}(I_{st})$ that in steady-state ensures that $z_2 = \lambda_{O_2}^d = 2$. It can be seen that the time needed for P_{net} to reach the desired value is approximately one sec.

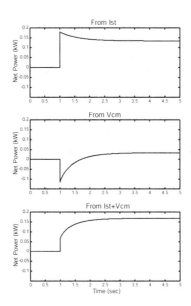

Fig. 5.25. Responses of P_{net} to steps in (i) I_{st}, (ii) v_{cm}, and (iii) coordinated I_{st} and v_{cm}

It is apparent that to speed up the P_{net} response, we need either a larger magnitude of I_{st} (to increase stack power) or a smaller value of v_{cm} (to decrease the parasitic losses). Either case will degrade the speed of the λ_{O_2} response because larger I_{st} causes additional drops in λ_{O_2} and smaller v_{cm} slows down the recovery rate of λ_{O_2}. The tradeoff between P_{net} and λ_{O_2} responses cannot be eliminated because there is only one control actuator. The actuator has to compromise between the two conflicting performance variables.

We systematically explore the tradeoff by setting up the LQ control problem with the cost function in terms of both performance variables:

$$J = \int_0^\infty Q_{z1}z_1^2 + Q_{z2}z_2^2 + Ru^2 + Q_I q^2 \, dt \qquad (5.31)$$

Figure 5.26 shows the time responses of the linear model with the different control gains based on different weighting in the cost function. The tradeoff between P_{net} and λ_{O_2} is evident during transient. In particular, controller design 4 (solid line) corresponds to the best power response but at the expense of slow recovery of the excess oxygen ratio. On the other hand, the fast recovery of excess oxygen ratio (dotted line) causes a net power lag of 0.200 sec which might be considered undesirable.

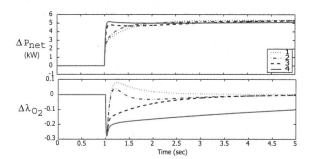

Fig. 5.26. Linear system response: 1 – feedforward 2 3 4 feedback with different gains

The same results are shown in the frequency domain using the Bode magnitude plots of Figure 5.27. The closer the two graphs are to zero, the better regulation is achieved. It can be seen that there is a severe tradeoff between the two variables in the frequency range between 0.7 rad/sec and 20 rad/sec. At these frequencies, when the magnitude of the upper variable is pushed closer to zero, the magnitude of the lower variable "pops up" indicating worse λ_{O_2} regulation. To decide on the best compromise between the two performance objectives, one needs to first establish a measure of how important to the stack life are the deviations in the excess oxygen ratio.

One option to overcome the tradeoff is to filter the current drawn from the stack and to use an additional energy storage device (battery or ultra-

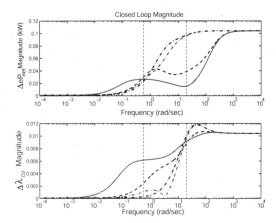

Fig. 5.27. Closed loop frequency responses for different control gains.

capacitor) to supplement the system power during transient. Another option is to have an oxygen storage device placed near the entrance of the stack to provide an instant oxygen supply during rapid current changes. The required size of the energy or oxygen storage devices could be determined based on the frequencies associated with the tradeoff (Figures 5.13 and 5.27). The control analysis with the dynamics model of the fuel cell system provides an important tool to identify the required sizes of these storage devices. Without this control analysis, it is very likely that unnecessary weight and volume would be added to the FCS system by oversized auxiliary components.

6

Natural Gas Fuel Processor System Model

A PEM fuel cell system that is not fueled by pure hydrogen can use a fuel processor to convert its primary fuel into hydrogen. For residential applications, fueling the fuel cell system using natural gas is often preferred because of its wide availability and extended distribution system [36]. Common methods of converting natural gas to hydrogen include steam reforming and partial oxidation. The most common method, steam reforming, which is endothermic, is well suited for steady-state operation and can deliver a relatively high concentration of hydrogen [2], but it suffers from poor transient response [23]. On the other hand, the partial oxidation offers several other advantages such as compactness, rapid start-up, and responsiveness to load changes [36], but delivers lower conversion efficiency.

The main reactor of a partial oxidation-based natural gas fuel processing system (FPS) is a catalytic partial oxidation (CPOX) reactor. Here, hydrogen-rich gas is produced by mixing natural gas with air over a catalyst bed. The amount of hydrogen created in the FPS depends on both the catalyst bed temperature and the CPOX air-to-fuel ratio, more specifically, the oxygen-to-carbon ratio. This oxygen-to-carbon ratio also influences the amount of heat generated in the CPOX, which then affects the CPOX catalyst bed temperature.

System-level dynamic models of fuel cell power plants built from physics-based component models are extremely useful in understanding the system-level interactions, implications for system performance, and model-aided controller design. The system-level dynamic models also help in evaluating alternative system architectures in an integrated design and control paradigm. In this chapter, we develop a dynamic model for the FPS control of the air blower and the fuel valve for fast and efficient H_2 generation. The FPS model is developed with a focus on the dynamic behavior associated with the flow and pressure in the various FPS reactor stages and the temperature of the CPOX. We neglect variations of the pressure, concentration, and temperature and lump them into spatially averaged variables that can be described using ordinary differential equations. The model is parameterized and validated against simulation results from a high-order fuel cell system model [40].

6.1 Fuel Processing System (FPS)

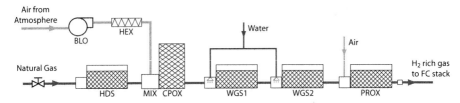

Fig. 6.1. FPS components

Figure 6.1 illustrates the components in a natural gas fuel processing system [111]. The FPS is composed of four main reactors, namely, hydro-desulfurizer (HDS), catalytic partial oxidation (CPOX), water gas shift (WGS), and preferential oxidation (PROX). Natural gas (Methane CH_4) is supplied to the FPS from either a high-pressure tank or a high-pressure pipeline. Sulfur, which poisons the water gas shift catalyst [23], is then removed from the natural gas stream in the HDS [36, 49]. The main air flow is supplied to the system by a blower (BLO) which draws air from the atmosphere. The air is then heated in the heat exchanger (HEX). The heated air and the de-sulfurized natural gas stream are then mixed in the mixer (MIX). The mixture is then passed through the catalyst bed inside the catalytic partial oxidizer where CH_4 reacts with oxygen to produce H_2. There are two main chemical reactions taking place in the CPOX, namely, partial oxidation (POX) and total oxidation (TOX) [122, 68]:

$$(POX) \quad CH_4 + \frac{1}{2}O_2 \rightarrow CO + 2H_2 \quad \Delta H^0_{pox} = -0.036 \times 10^6 \text{ J/mol} \quad (6.1)$$

$$(TOX) \quad CH_4 + 2O_2 \rightarrow CO_2 + 2H_2O \quad \Delta H^0_{tox} = -0.8026 \times 10^6 \text{ J/mol} \quad (6.2)$$

Heat is released from both reactions. However, the TOX reaction releases much more heat than the POX reaction. The rates of the two reactions depend on the selectivity S, defined as

$$S = \frac{\text{rate of } CH_4 \text{ reacting in POX}}{\text{total rate of } CH_4 \text{ reacting}} \quad (6.3)$$

The selectivity depends strongly on the oxygen-to-carbon (O2C) ratio (O_2 to CH_4) entering the CPOX [122]. Hydrogen is created only in the POX reaction and, therefore, it is preferable to promote this reaction in the CPOX. However, carbon monoxide (CO) is created along with H_2 in the POX reaction as can be seen in (6.1). Because CO poisons the fuel cell catalyst, it is eliminated using both the water gas shift converter and the preferential oxidizer. As illustrated in Figure 6.1, there are typically two WGS reactors operating at different temperatures [23, 71]. In the WGS, water is injected into the gas flow to promote a water gas shift reaction:

$$(WGS) \qquad CO + H_2O \rightarrow CO_2 + H_2 \qquad (6.4)$$

Note that even though the objective of WGS is to eliminate CO, hydrogen is also created from the WGS reaction. The level of CO in the gas stream after WGS is still high for fuel cell operations and thus oxygen is injected (in the form of air) into the PROX reactor to react with the remaining CO:

$$(PROX) \qquad 2CO + O_2 \rightarrow 2CO_2 \qquad (6.5)$$

The amount of air injected into the PROX is typically twice the amount that is needed to maintain the stoichiometric reaction in (6.5) [23, 37].

There are two main control objectives. First, to prevent stack H_2 starvation [120, 102], which can permanently damage the stack, the hydrogen flow from the FPS must respond rapidly and robustly to changes in the stack power level (*i.e.*, changes in stack current [93]). Unfortunately, oversupply of H_2 by adjusting the FPS flow at a higher steady-state level is not an option because this will cause wasted hydrogen from the anode exhaust [102]. Thus, hydrogen generation needs to follow the current load in a precise and fast manner.

Second, the temperature of the CPOX must be maintained at a certain level. Exposure to high temperature will permanently damage the CPOX catalyst bed whereas low CPOX temperature reduces the reaction rate [122]. The optimization of these goals during transient operations can be achieved by coordinating the CPOX air blower command and the fuel (natural gas) valve command.

6.2 Control-oriented FPS Model

The FPS model is developed with a focus on the dynamic behaviors associated with the flows and pressures in the FPS and also the temperature of the CPOX. The dynamic model is used to study the effects of fuel and air flow command to (i) CPOX temperature [122], (ii) stack H_2 concentration [102], and (iii) steady-state stack efficiency. The stack efficiency is interpreted as the H_2 utilization, which is the ratio between the hydrogen reacted in the fuel cell stack and the amount of hydrogen supplied to the stack.

6.2.1 Modeling Assumptions

Several assumptions are made to simplify the FPS model. Because the control of WGS and PROX reactants is not considered in this study, the two components are lumped together as one volume and the combined volume is called WROX (WGS+PROX). It is assumed that both components are perfectly controlled such that the desired values of the reactants are supplied to the reactors. Furthermore, because the amount of H_2 created in WGS is proportional to the amount of CO eliminated in WGS (Reaction (6.4)), which in turn is proportional to the amount of H_2 generated in CPOX (Reaction (6.1)), it is assumed that the amount of H_2 generated in the WGS is always a fixed percentage of the amount of H_2 produced in the CPOX. The desulfurization process in the HDS is not modeled and thus the HDS is viewed as a storage volume. It is assumed that the composition of the air entering the blower is constant. Additionally, all temperatures except the CPOX temperature are assumed constant and the effect of temperature changes on the pressure dynamics is assumed negligible. The volume of CPOX is relatively small and is ignored. It is also assumed that the CPOX reaction is rapid and reaches equilibrium before the flow exits the CPOX reactor. Finally, all gases obey the ideal gas law and all gas mixtures are perfect mixtures. Figure 6.2 illustrates the simplified system and state variables used in the model. The physical constants used throughout the model are given in Table 6.1 and the properties of the air entering the blower (approximately 40% relative humidity) are given in Table 6.2.

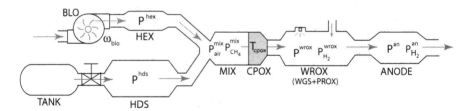

Fig. 6.2. FPS dynamic model

Table 6.1. Physical constants

Parameter	Value
R	8.3145 J/mol·K
M_{N_2}	28×10^{-3} kg/mol
M_{CH_4}	16×10^{-3} kg/mol
M_{CO}	28×10^{-3} kg/mol
M_{CO_2}	44×10^{-3} kg/mol
M_{H_2}	2×10^{-3} kg/mol
M_{H_2O}	18×10^{-3} kg/mol
M_{O_2}	32×10^{-3} kg/mol
F	96,485 Coulombs

Table 6.2. Conditions of the atmospheric air entering the blower

Parameter	Value
p_{amb}	1×10^5 Pa
$y_{N_2}^{atm}$	0.6873
$y_{H_2O}^{atm}$	0.13
$y_{O_2}^{atm}$	0.1827
M_{air}^{atm}	27.4×10^{-3} kg/mol

6.2.2 Model States and Principles

The dynamic states in the model, shown in Figure 6.2, are blower speed ω_{blo}, heat exchanger pressure p^{hex}, HDS pressure p^{hds}, mixer CH_4 partial pressure $p_{CH_4}^{mix}$, mixer air partial pressure p_{air}^{mix}, CPOX temperature T_{cpox}, WROX (combined WGS and PROX) volume pressure p^{wrox}, WROX hydrogen partial pressure $p_{H_2}^{wrox}$, anode pressure p^{an}, and anode hydrogen partial pressure $p_{H_2}^{an}$. Mass conservation with the ideal gas law through the isothermal assumption is used to model the filling dynamics of the gas in all volumes considered in the system. The orifice equation with a turbulent flow assumption is used to calculate flow rates between two volumes. The energy conservation principle is used to model the changes in CPOX temperature. The conversion of the gases in CPOX is based on the reactions in (6.1) and (6.2) and the selectivity defined in (6.3).

6.2.3 Orifice

The flow between any two volumes in the FPS system is based on the orifice flow equation. Specifically, the mass flow rate between two volumes is given as a function of upstream pressure p_1 and downstream pressure p_2. The flow is assumed turbulent and the rate is governed by

$$W = W_0 \sqrt{\frac{p_1 - p_2}{\Delta p_0}} \tag{6.6}$$

where W_0 and Δp_0 are the nominal air flow rate and the nominal pressure drop of the orifice, respectively.

6.2.4 Blower (BLO)

The speed of the blower is modeled as a first-order dynamic system with time constant τ_b. The governing equation is

$$\frac{d\omega_{blo}}{dt} = \frac{1}{\tau_b}\left(\frac{u_{blo}}{100}\omega_0 - \omega_{blo}\right) \tag{6.7}$$

where u_{blo} is the blower command signal (range between 0 and 100) and ω_0 is the nominal blower speed (3600 rpm). The gas flow rate through the blower W_{blo} is determined using the blower map, which represents the relation between a scaled blower volumetric flow rate and a scaled pressure head [21]. The scaled pressure head is the actual pressure head scaled by a square of the speed ratio; that is,

$$[\text{scaled pressure head}] = [\text{actual head}] \left(\frac{\omega}{\omega_0} \right)^2 \qquad (6.8)$$

and the scaled volumetric flow rate is the actual flow rate scaled by the reciprocal of the speed ratio; that is,

$$[\text{scaled flow}] = \frac{[\text{actual flow}]}{\left(\frac{\omega}{\omega_0} \right)} \qquad (6.9)$$

Note that the changes in gas density are ignored and thus only the blower speed is used in the scaling. The blower mass flow rate W_{blo} is calculated by multiplying the volumetric flow rate with constant air density (1.13 kg/m^3). The blower map is shown in Figure 6.3 and the blower time constant is 0.3 sec.

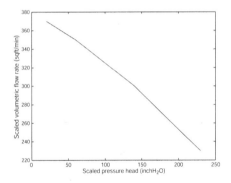

Fig. 6.3. Blower map

6.2.5 Heat Exchanger Volume (HEX)

The only dynamics considered in the heat exchanger is the pressure dynamics. The changes in temperature of the gas are ignored and it is assumed that the effects of actual temperature changes on the pressure dynamics are negligible. The rate of change in air pressure of the HEX is described by

$$\frac{dp^{hex}}{dt} = \frac{RT_{hex}}{M_{air}^{atm}V_{hex}} (W^{blo} - W^{hex}) \qquad (6.10)$$

where M_{air}^{atm} is the molecular weight of the air flow through the blower (given in Table 6.2). The orifice flow equation (6.6) is used to calculate the outlet flow rate of the HEX W^{hex}, as a function of HEX pressure p^{hex}, and mixer pressure p^{mix}.

6.2.6 Hydro-desulfurizer Volume (HDS)

The pressure of the gas in the HDS is governed by the mass balance principle. It is assumed that the natural gas fed to the HDS is pure methane (CH_4) [23], and thus the desulfurization process is not modeled. The HDS is then considered as a gas volume and the pressure changes are modeled by

$$\frac{dp^{hds}}{dt} = \frac{RT_{hds}}{M_{CH_4}V_{hds}}(W_{fuel} - W^{hds}) \tag{6.11}$$

where W^{hds} is the rate of mass flow from HDS to the mixer (MIX), and is calculated as a function of p^{hds} and p^{mix} using the orifice equation (6.6). The temperature of the gas T_{hds} is assumed constant.

The flow rate of methane into the HDS W_{fuel} is controlled by a fuel valve. The orifice equation (6.6) with variable gain based on the valve input signal u_{valve} (0 to 100) is used to model the flow through the valve.

$$W_{fuel} = \left(\frac{u_{valve}}{100}\right) W_{0,valve} \sqrt{\frac{p^{tank} - p^{hds}}{\Delta p_{0,valve}}} \tag{6.12}$$

where p^{tank} is the fuel tank or supply line pressure.

6.2.7 Mixer (MIX)

The natural gas flow from the HDS W^{hds}, and the air flow from the blower W^{hex}, are combined in the mixer. Two dynamic variables in the mixer model are the methane pressure $p_{CH_4}^{mix}$ and the air pressure p_{air}^{mix}. The state equations of the MIX model are

$$\frac{dp_{CH_4}^{mix}}{dt} = \frac{RT_{mix}}{M_{CH_4}V_{mix}}(W^{hds} - x_{CH_4}^{mix}W^{cpox}) \tag{6.13}$$

$$\frac{dp_{air}^{mix}}{dt} = \frac{RT_{mix}}{M_{air}^{atm}V_{mix}}(W^{hex} - x_{air}^{mix}W^{cpox}) \tag{6.14}$$

where W^{cpox} is the flow rate through the CPOX which is calculated in Section 6.2.8. The mixer total pressure is the sum of the CH_4 and the air pressures $p^{mix} = p_{CH_4}^{mix} + p_{air}^{mix}$. Based on $p_{CH_4}^{mix}$ and p_{air}^{mix}, the mass fractions of CH_4 and the air in the mixer $x_{CH_4}^{mix}$ and x_{air}^{mix} are calculated by

$$x^{mix}_{CH_4} = \cfrac{1}{1 + \cfrac{M^{atm}_{air}}{M_{CH_4}} \cfrac{p^{mix}_{air}}{p^{mix}_{CH_4}}} \qquad (6.15)$$

$$x^{mix}_{air} = \cfrac{1}{1 + \cfrac{M_{CH_4}}{M^{atm}_{air}} \cfrac{p^{mix}_{CH_4}}{p^{mix}_{air}}} \qquad (6.16)$$

where M_{CH_4} and M^{atm}_{air} are the molar masses of methane and atmospheric air, respectively (see Table 6.2). Note that $x^{mix}_{CH_4} + x^{mix}_{air} = 1$ because the gas in MIX volume is composed only of methane and atmospheric air. The temperature of the mixer gas T_{mix} is assumed constant.

The mass fractions of nitrogen, oxygen, and vapor in the mixer needed for the calculation of the CPOX reactions are calculated by

$$x^{mix}_{N_2} = x^{atm}_{N_2} x^{mix}_{air} \qquad (6.17)$$

$$x^{mix}_{O_2} = x^{atm}_{O_2} x^{mix}_{air} \qquad (6.18)$$

$$x^{mix}_{H_2O} = x^{atm}_{H_2O} x^{mix}_{air} \qquad (6.19)$$

where x^{atm}_i is the mass fraction of species i in atmospheric air, which is calculated from the mole fractions given in Table 6.2. Note that $x^{mix}_{N_2} + x^{mix}_{O_2} + x^{mix}_{H_2O} = x^{mix}_{air}$. The oxygen-to-carbon, that is, O_2-to-CH_4, (mole) ratio λ_{O2C}, which influences the reaction rate in the CPOX, is calculated by

$$\lambda_{O2C} \equiv \frac{n_{O_2}}{n_{CH_4}} = y^{atm}_{O_2} \frac{p^{mix}_{air}}{p^{mix}_{CH_4}} \qquad (6.20)$$

where n_i is the number of moles of species i, and $y^{atm}_{O_2}$ is the oxygen mole fraction of the atmospheric air.

6.2.8 Catalytic Partial Oxidation (CPOX)

Because the gas volume in the CPOX catalyst bed is relatively small, the pressure dynamics of the gas is ignored. The flow rate though the CPOX W^{cpox} is calculated using the orifice equation (6.6) as a function of mixer total pressure p^{mix} and the total pressure in WGS and PROX combined volume p^{wrox}. The only dynamics considered in the CPOX is the catalyst temperature T_{cpox}. The temperature dynamics is modeled using the energy balance equation

$$m^{cpox}_{bed} C^{cpox}_{P,bed} \frac{dT_{cpox}}{dt} = \begin{bmatrix} \text{inlet} \\ \text{enthalpy} \\ \text{flow} \end{bmatrix} - \begin{bmatrix} \text{outlet} \\ \text{enthalpy} \\ \text{flow} \end{bmatrix} + \begin{bmatrix} \text{heat from} \\ \text{reactions} \end{bmatrix} \quad (6.21)$$

where m^{cpox}_{bed} (kg) and $C^{cpox}_{P,bed}$ (J/kg·K) are the mass and specific heat capacity of the catalyst bed, respectively. The last two terms on the right-hand side of (6.21) depend on the reaction taking place in the CPOX.

In the catalytic partial oxidation reactor, methane CH_4 is oxidized to produce hydrogen. There are two CH_4 oxidation reactions: partial oxidation and total oxidation.

$(POX) \quad CH_4 + \frac{1}{2}O_2 \rightarrow CO + 2H_2$

$$\Delta H^0_{pox} = -0.036 \times 10^6 \text{ J/mol of } CH_4 \qquad (6.22)$$

$(TOX) \quad CH_4 + 2O_2 \rightarrow CO_2 + 2H_2O$

$$\Delta H^0_{tox} = -0.8026 \times 10^6 \text{ J/mol of } CH_4 \qquad (6.23)$$

The other two secondary reactions considered here are water formation, or hydrogen oxidation (HOX), and carbon monoxide preferential oxidation (COX).

$(HOX) \quad 2H_2 + O_2 \rightarrow 2H_2O \quad \Delta H^0_{hox} = -0.4836 \times 10^6 \text{ J/mol of } O_2 \quad (6.24)$

$(COX) \quad 2CO + O_2 \rightarrow 2CO_2 \quad \Delta H^0_{cox} = -0.566 \times 10^6 \text{ J/mol of } O_2 \quad (6.25)$

The species entering the CPOX include CH_4, O_2, H_2O, and N_2. Nitrogen does not react in the CPOX. The water may react with CH_4 through steam reforming reaction; however, this reaction is ignored in this study. Methane reacts with oxygen to create the final product, which contains H_2, H_2O, CO, CO_2, CH_4, and O_2 [122]. The amount of each species depends on the initial oxygen-to-carbon (O_2 to CH_4) ratio λ_{o2c} of the reactants and the temperature of the CPOX catalyst bed T_{cpox}. All reactions in the CPOX occur concurrently. However, to simplify the model, we view the overall CPOX reaction as a sequential process of reactions (6.22) to (6.25), as illustrated in Figure 6.4.

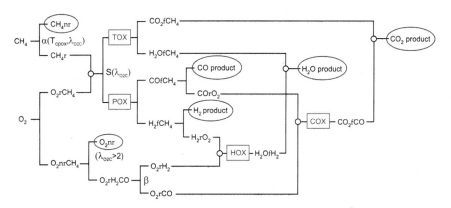

Fig. 6.4. Illustration of calculation of CPOX reactions

The figure notations are: r = "react," nr = "not react," and f = "from." Following the diagram, first, consider the CH_4 and O_2 that enter the CPOX. The amount of CH_4 that reacts is a function of both the O2C ratio and CPOX temperature. The relation is determined from the thermodynamic equilibrium analysis that was presented in [122]. Here, the relation is modeled using the variable α, defined as

$$\alpha := \frac{\text{rate of } CH_4 \text{ reacts}}{\text{rate of } CH_4 \text{ enters}} := \frac{N_{CH_4 r}}{N_{CH_4 in}} \qquad (6.26)$$

The expression of α is developed by curve fitting the results in [122].

$$\alpha = \begin{cases} \alpha_1 \lambda_{O2C} & \lambda_{O2C} < 0.5 \\ 1 - (1 - 0.5\alpha_1)(1 - \tanh(\alpha_2(\lambda_{O2C} - 0.5))) & \lambda_{O2C} \geq 0.5 \end{cases} \qquad (6.27)$$

where

$$\alpha_1 = \min(2, \, 0.0029 T_{cpox} - 1.185) \qquad (6.28)$$

$$\alpha_2 = 0.215 e^{3.9 \times 10^{-8}(T_{cpox} - 600)^3} \qquad (6.29)$$

For illustration purposes, Figure 6.5 shows a plot of $(1 - \alpha)$, which represents the amount of CH_4 that slips through the reactor (*i.e.*, does not react) as a function of λ_{O2C} and temperature T_{cpox}. For λ_{O2C} less than 0.5, the oxygen supplied is not enough to react with all the CH_4 and thus there is unreacted CH_4 left regardless of the CPOX temperature. For λ_{O2C} larger than 0.5, all CH_4 reacts for CPOX temperature over 1073 K. For lower temperature, not all CH_4 reacts, which means that part of the fuel is wasted. Note that the curve fitting does not fit well for lower temperature ($T_{cpox} < 700$ K) when compared with the results in [122]. However, as shown in Section 6.3, the FPS model is operated at CPOX temperature around 900 K to 1000 K where the model fits very well.

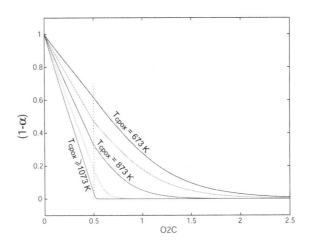

Fig. 6.5. Amount of unreacted CH_4 that leaves the CPOX

CH_4 engages in either POX or TOX reactions depending on the initial O2C ratio, which, in this model, is the O2C ratio in the MIX. The POX and

TOX reaction rates are determined by the selectivity S defined in (6.3),

$$S := \frac{N_{CH_4rPOX}}{N_{CH_4r}} \tag{6.30}$$

which is a function of λ_{o2c}. Here we assume that the function is linear, as shown in Figure 6.6, which agrees with the results from the high-temperature thermodynamic equilibrium in [122]. The relation between the selectivity and the oxygen-to-carbon ratio in Figure 6.6 can be expressed as

$$S = \begin{cases} 1 & , \lambda_{o2c} < \frac{1}{2} \\ \frac{2}{3}(2 - \lambda_{o2c}) & , \frac{1}{2} \leq \lambda_{o2c} \leq 2 \\ 0 & , \lambda_{o2c} > 2 \end{cases} \tag{6.31}$$

Larger values of S indicate that more CH_4 engages in the POX reaction and thus more hydrogen is generated. The products from CH_4 oxidation (POX and TOX) are H_2, CO, H_2O, and CO_2, denoted in Figure 6.4 as H_2fCH_4, $COfCH_4$, H_2OfCH_4, and CO_2fCH_4, respectively.

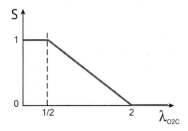

Fig. 6.6. Selectivity between POX and TOX

As explained earlier, when $\lambda_{o2c} < \frac{1}{2}$, the supplied oxygen is not sufficient to oxidize all supplied fuel and the hydrogen production rate is limited by the amount of oxygen. At normal operation, λ_{o2c} is kept higher than $\frac{1}{2}$ in order to avoid fuel waste. A high value of λ_{o2c} (low S) indicates that there is more TOX reaction. Because more heat is released from TOX reaction, operating CPOX at high λ_{o2c} will overheat the CPOX and can permanently damage the catalyst bed. The desired value of λ_{o2c} in the literature varies from 0.4 to 0.6 [25, 91, 95]. In this study, the desired value is chosen at $\lambda_{o2c} = 0.6$ to allow some buffer for λ_{o2c} before it becomes lower than $\frac{1}{2}$ during transient deviations.

The amount of O_2 that does not react (nr) with CH_4 (O_2nrCH_4), reacts with H_2 and CO, created in the POX reaction, to form H_2O (HOX reaction) and CO_2 (COX reaction), respectively. If there are no H_2 and CO generated (no POX reaction), there will be unreacted O_2 (O_2nr), which then leaves the CPOX. This corresponds to the situation where $\lambda_{o2c} \geq 2$. If $\lambda_{o2c} < 2$, all O_2 that does not react with CH_4 will react with H_2 and CO (O_2rH_2CO). In this model, the rate of HOX and COX reactions is described by the variable β, defined as

$$\beta := \frac{\text{rate of } O_2 \text{ reacts with } H_2}{\text{rate of } O_2 \text{ reacts with both } H_2 \text{ and CO}} := \frac{N_{O_2rH_2}}{N_{O_2rH_2CO}} \qquad (6.32)$$

In POX, there are two moles of H_2 produced per one mole of CO produced and O_2 reacts with H_2 more than CO. Therefore, the ratio β is kept constant at $\beta = \frac{2}{3}$. The water product of the HOX reaction, denoted as H_2OfH_2 (water from H_2) is then added to the water produced in the TOX reaction. Similarly, the final product of CO_2 is the sum of CO_2 from the TOX reaction (CO_2fCH_4) and CO_2 from the COX reaction (CO_2fCO). The final products of H_2 and CO are the amount produced in the POX reaction (H_2fCH_4 and $COfCH_4$) less the amount that reacts with O_2 (H_2rO_2 and $COrO_2$), which can be easily calculated using stoichiometry of the HOX and COX reactions.

The species in the CPOX model are calculated in mole basis. The molar flow rate of the gas entering the CPOX can be calculated from

$$N_{i,in} = \frac{x_i^{mix}W_{cpox}}{M_i} \qquad (6.33)$$

where i represents CH_4, O_2, N_2, and H_2O; M_i is the molecular mass of gas i; and W_{cpox} and x_i^{mix} are the CPOX total flow rate and mole fraction of gas i in MIX; both are calculated in the MIX model (Equations (6.15) to (6.19)). From the definition of α, the rate at which CH_4 reacts is

$$N_{CH_4r} = \alpha N_{CH_4in} \qquad (6.34)$$

and the rate at which O_2 reacts with CH_4 is

$$N_{O_2rCH_4} = (2 - \frac{3}{2}S)N_{CH_4r} = (2 - \frac{3}{2}S)\alpha N_{CH_4in} \qquad (6.35)$$

Thus, the rate of O_2 not reacting with CH_4 is

$$N_{O_2nrCH_4} = N_{O_2in} - (2 - \frac{3}{2}S)\alpha N_{CH_4in} = (\lambda_{o2c} - (2 - \frac{3}{2}S)\alpha)N_{CH_4in} \qquad (6.36)$$

If there is a POX reaction ($S \neq 0$), the oxygen that does not react with CH_4 will either react with H_2 or CO. If there is no POX reaction ($S = 0$), there is no H_2 or CO to react with the oxygen. The amount of O_2 that reacts either with H_2 or CO ($N_{O_2rH_2CO}$) and the amount of unreacted O_2 (N_{O_2nr}) are

$$N_{O_2rH_2CO} = N_{O_2nrCH_4}\text{sign}(S)$$
$$= (\lambda_{o2c} - (2 - \frac{3}{2}S)\alpha)\text{sign}(S)N_{CH_4in} \qquad (6.37)$$
$$N_{O_2nr} = N_{O_2nrCH_4}(1 - \text{sign}(S))$$
$$= (\lambda_{o2c} - (2 - \frac{3}{2}S)\alpha)(1 - \text{sign}(S))N_{CH_4in} \qquad (6.38)$$

The product of H_2, CO, CO_2, and H_2O from POX and TOX reactions can be calculated from

$$N_{H_2fCH_4} = 2S \cdot N_{CH_4r} = 2S \cdot \alpha N_{CH_4in} \tag{6.39}$$

$$N_{COfCH_4} = S \cdot N_{CH_4r} = S \cdot \alpha N_{CH_4in} \tag{6.40}$$

$$N_{CO_2fCH_4} = (1 - S) \cdot N_{CH_4r} = (1 - S) \cdot \alpha N_{CH_4in} \tag{6.41}$$

$$N_{H_2OfCH_4} = 2(1 - S) \cdot N_{CH_4r} = 2(1 - S) \cdot \alpha N_{CH_4in} \tag{6.42}$$

The rate of H_2 and CO reacted and the rate of H_2O and CO_2 created in HOX and COX reactions are

$$N_{H_2rO_2} = 2\beta \cdot N_{O_2rH_2CO} \tag{6.43}$$

$$N_{COrO_2} = 2(1 - \beta) \cdot N_{O_2rH_2CO} \tag{6.44}$$

$$N_{CO_2fCO} = 2(1 - \beta) \cdot N_{O_2rH_2CO} \tag{6.45}$$

$$N_{H_2OfH_2} = 2\beta \cdot N_{O_2rH_2CO} \tag{6.46}$$

Combining Equations (6.37), (6.42), and (6.46), a set of equations to calculate the total products of CPOX reactions can be written as

$$N_{H_2} = N_{H_2fCH_4} - N_{H_2rO_2}$$
$$= \left[2S\alpha - 2\beta(\lambda_{o2c} - (2 - \frac{3}{2}S)\alpha)\text{sign}(S) \right] N_{CH_4in} \tag{6.47a}$$

$$N_{CO} = N_{COfCH_4} - N_{COrO_2}$$
$$= \left[S\alpha - 2(1 - \beta)(\lambda_{o2c} - (2 - \frac{3}{2}S)\alpha)\text{sign}(S) \right] N_{CH_4in} \tag{6.47b}$$

$$N_{CO_2} = N_{CO_2fCH_4} + N_{CO_2fCO}$$
$$= \left[(1 - S)\alpha + 2(1 - \beta)(\lambda_{o2c} - (2 - \frac{3}{2}S)\alpha)\text{sign}(S) \right] N_{CH_4in} \tag{6.47c}$$

$$N_{H_2O} = N_{H_2OfCH_4} + N_{H_2OfH_2} + N_{H_2Oin}$$
$$= \left[2(1 - S)\alpha + 2\beta(\lambda_{o2c} - (2 - \frac{3}{2}S)\alpha)\text{sign}(S) \right] N_{CH_4in} + N_{H_2Oin} \tag{6.47d}$$

$$N_{CH_4} = (1 - \alpha)N_{CH_4in} \tag{6.47e}$$

$$N_{O_2} = N_{O_2in} - N_{O_2r} = \left(N_{O_2in} - (2 - \frac{3}{2}S)\alpha N_{CH_4in} \right)\text{sign}(S) \tag{6.47f}$$

$$N_{N_2} = N_{N_2in} \tag{6.47g}$$

A plot of products calculated from (6.47), assuming no inlet N_2 and H_2O, is shown in Figure 6.7, which matches with the theoretical results in [122]. The mass flow rate of each species leaving the CPOX is $W_i^{cpox} = M_i N_i$. The mass conservation property of chemical reactions ensures that the total mass flow across the CPOX is conserved; that is, $\sum_i W_i^{cpox} = W^{cpox}$.

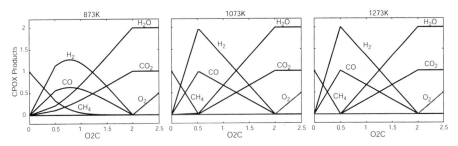

Fig. 6.7. Products of CPOX reaction per unit of CH_4 entering CPOX

The dynamic equation of temperature (6.21) can now be expanded. The enthalpy of the gas flow depends on the flow rate, the flow temperature, and the gas composition. Thus

$$\left[\begin{array}{c} \text{Enthalpy flow} \\ \text{in - out} \end{array}\right] = W^{cpox}\left(C_P^{mix}(T_{mix} - T_{ref}) - C_P^{cpox}(T_{cpox} - T_{ref})\right)$$
(6.48)

where T_{ref} is the reference temperature (298 K). The gas specific heat C_P^{mix} and C_P^{cpox} $(J/kg \cdot K)$ are that of the gas in the mixer (gas before the CPOX reaction) and the gas in the CPOX (after reaction), respectively. They are functions of gas composition and gas temperature.

$$C_P^{mix} = \sum x_i^{mix} C_{P_i}(T_{mix})$$
(6.49)

$$C_P^{cpox} = \sum x_i^{cpox} C_{P_i}(T_{cpox})$$
(6.50)

where i represents four species in the MIX (Equation (6.33)) and seven species in the CPOX (Equation (6.47)). The heat released from the reaction depends on the amount of reaction taking place.

$$\left[\begin{array}{c} \text{Heat from} \\ \text{reaction} \end{array}\right] = N_{CH_4 r}\left(S \cdot (-\Delta H_{pox}^0) + (1 - S) \cdot (-\Delta H_{tox}^0)\right)$$
$$+ N_{O_2 r H_2 CO}\left(\beta \cdot (-\Delta H_{hox}^0) + (1 - \beta) \cdot (-\Delta H_{cox}^0)\right) \text{ (6.51)}$$

where $-\Delta H_{pox}^0$, $-\Delta H_{tox}^0$, $-\Delta H_{hox}^0$, and $-\Delta H_{cox}^0$ (J/mol), given in Equations (6.22) to (6.25), are the heat released from the POX, TOX, HOX, and COX reactions, respectively. Combining Equations (6.21), (6.48), and (6.51), the state equation of the CPOX temperature can be written as

$$\frac{dT_{cpox}}{dt} = \frac{1}{m_{bed}^{cpox} C_{P,bed}^{cpox}}\left[W^{cpox}\left(C_P^{mix}(T_{mix} - T_{ref}) - C_P^{cpox}(T_{cpox} - T_{ref})\right) + \right.$$
$$N_{CH_4 r}\left(S \cdot (-\Delta H_{pox}^0) + (1 - S) \cdot (-\Delta H_{tox}^0)\right)$$
$$\left. + N_{O_2 r H_2 CO}\left(\beta \cdot (-\Delta H_{hox}^0) + (1 - \beta) \cdot (-\Delta H_{cox}^0)\right)\right]$$
(6.52)

6.2.9 Water Gas Shift Converter and Preferential Oxidation Reactor (WROX)

The water gas shift converter and the preferential oxidation reactor are lumped together as one volume, denoted as WROX. The three flows entering the volume are H_2-rich gas flow from the CPOX W^{cpox}, water injection needed for WGS reaction $W^{wgs}_{H_2O}$, and air injection required for PROX reaction W^{prox}_{air}. The flow rates of water injected into WGSs are equal to the amount that is needed to cool down the gas temperature to the desired WGS inlet temperatures [23, 37]. The amount of air supplied to the PROX reactor is normally twice that required to oxidize the rest of the CO in the gas stream based on the desired operating condition [23, 37]. The WROX model has two states: total pressure p^{wrox} and hydrogen pressure $p^{wrox}_{H_2}$. Because the amount of CO created in CPOX is proportional to the rate of H_2 created (POX reaction), it is assumed, in the WROX model, that the rate of H_2 generated in the WGS is a fixed percentage (η_{wrox}) of the rate of hydrogen generated in the CPOX. The state equations are

$$\frac{dp^{wrox}}{dt} = \frac{RT_{wrox}}{M_{wrox}V_{wrox}} \left(W^{cpox} - W^{wrox} + W^{wgs}_{H_2O} + W^{prox}_{air} \right) \quad (6.53)$$

$$\frac{dp^{wrox}_{H_2}}{dt} = \frac{RT_{wrox}}{M_{H_2}V_{wrox}} \left((1 + \eta_{wrox})W^{cpox}_{H_2} - x^{wrox}_{H_2}W^{wrox} \right) \quad (6.54)$$

where M_{wrox} is an average molecular weight of the gas in WROX, and T_{wrox} is an average temperature of WGSs and PROX. The WROX exit flow rate W^{wrox} is calculated using the nozzle equation (6.6) based on the pressure drop between WROX and anode volume $p^{wrox} - p^{an}$. The hydrogen mass fraction in the WROX $x^{wrox}_{H_2}$ can be determined from the two states by

$$x^{wrox}_{H_2} = \frac{M_{H_2}}{M_{wrox}} \frac{p^{wrox}_{H_2}}{p^{wrox}} \quad (6.55)$$

The rate of water injected into WROX $W^{wgs}_{H_2O}$ is equal to the amount required to cool the gas from CPOX down to the desired WGSs inlet temperatures. There are two WGS reactors and thus the total rate of water injected is $W^{wgs}_{H_2O} = W^{wgs1}_{H_2O} + W^{wgs1}_{H_2O}$. The flow rate of water into each WGS is calculated using the energy balance between enthalpy of the gas flows, enthalpy of the flow at the desired temperature, and the heat of water vaporization. It is assumed that PROX air injection W^{prox}_{air} is scheduled based on the stack current at the value twice needed [23, 37] at the designed operating condition.

6.2.10 Anode (AN)

Mass conservation is used to model the pressure dynamic in the anode volume. To simplify the model, only three mass flows are considered, including flows into and out of the anode volume and the rate of hydrogen consumed in the fuel cell reaction. The dynamic equations are

$$\frac{dp^{an}}{dt} = \frac{RT_{an}}{M_{an}V_{an}} \left(W^{wrox} - W^{an} - W_{H_2,react} \right) \tag{6.56}$$

$$\frac{dp^{an}_{H_2}}{dt} = \frac{RT_{an}}{M_{H_2}V_{an}} \left(x^{wrox}_{H_2} W^{wrox} - x^{an}_{H_2} W^{an} - W_{H_2,react} \right) \tag{6.57}$$

where W^{an} is calculated as a function of the anode pressure p^{an} and the ambient pressure p_{amb} using Equation (6.6). The rate of hydrogen reacted is a function of stack current I_{st} through the electrochemistry principle [69]

$$W_{H_2,react} = M_{H_2} \frac{nI_{st}}{2F} \tag{6.58}$$

where n is the number of fuel cells in the stack and F is the Faraday number (given in Table 6.1).

The two important performance variables for the anode are the hydrogen utilization U_{H_2} and the anode hydrogen mole fraction y_{H_2}, which are calculated from

$$U_{H_2} = \frac{W_{H_2,react}}{x^{wrox}_{H_2} W^{wrox}} \tag{6.59}$$

and

$$y_{H_2} = \frac{p^{an}_{H_2}}{p^{an}} \tag{6.60}$$

The hydrogen utilization represents stack efficiency and the hydrogen mole fraction is used as an indication of stack H_2 starvation. Low hydrogen utilization means that more hydrogen is wasted through the anode exhaust and thus the stack efficiency is reduced. High utilization corresponds to high fuel cell efficiency, but it increases the risk of fuel cell H_2 starvation during transient.

6.3 Simulation and FPS Model Validation

The low-order (10 states) model described in Section 6.2 is developed in the MATLAB®/Simulink® platform. The model is parameterized and validated with the results of a high-order (>300 states) detailed model [40] developed using the Modelica language and Dymola™ software [114]. Here, we describe briefly the detailed physics-based Dymola™ model used for validating the simplified model presented in this chapter.

Besides the fuel processing system the Modelica fuel cell system model includes the cell stack assembly (CSA), power conditioning system (PCS), and the thermal management system (TMS) to accommodate the needs of several design teams in stationary and automotive applications. It consists of hundreds of components, several hundred dynamic states, and more than

20,000 equations. The FPS and CSA are the most complex subsystems. The FPS comprises detailed models of all the previously described reaction stages, namely, CPOX, two WGS, and two PROX reactors. The reactor models are described as continuous stirred tanks (CSTR) and employ bulk rate expressions based on experimental data. Besides the oxidation reactions for methane and hydrogen, carbon monoxide is also considered. The detailed model contains a complex flow network near ambient pressure with several recycle flows. The CSA cathode and anode volumes are discretized with multinode approximations and separated with membrane models that incorporate gas and liquid diffusion.

The focus of our work is to capture the essential dynamic input/output behavior associated with the hydrogen generation, and thus our main focus is in pressures, temperatures, and flows upstream from the CPOX reactor. We assume that temperature deviations in a well-controlled WROX subsystem do not propagate upstream in the CPOX reactor. The pressure dynamics of the WROX subsystem can, however, affect the flow through the CPOX and thus are captured in the control-oriented (10 states) model. Note here that the simplified model is based on well-controlled (constant and nominal) conditions in the WGS and the PROX that allow us to lump them in one volume equation (WROX). We test these important assumptions by augmenting the detailed Dymola model with several heat exchangers and a complex thermal management system in order to ensure good temperature conditions in all reactor stages and in the anode of the CSA.

The two models are compared with equivalent inputs after the Dymola model is imported as an S-function in Simulink®. The model parameters for a system designed to be used for residential or commercial buildings are given in Table 6.3. A similar power range would be needed for a bus or a heavy-duty vehicle propulsion system. The FPS key performance variables are the O2C ratio, the CPOX temperature, the FPS exit total flow rate, and the FPS exit hydrogen flow rate. Several parameters, such as the orifice constants and the component volumes, are adjusted appropriately in order to obtain comparable transient responses. Note that the model is expected to provide a close prediction of the transient response of the variables located upstream from the WGS inlet (WROX inlet). On the other hand, a relatively large discrepancy is expected for the variables downstream from the CPOX because the WGS and PROX reactors are approximately modeled as one lumped volume and are assumed perfectly controlled, which is not the case for the Dymola model.

The nominal operating point used in the validation is chosen at the oxygen-to-carbon ratio $\lambda_{O2C} = 0.6$ and the stack hydrogen utilization $U_{H_2} = 80\%$ [37]. The results are shown in Figures 6.8 to 6.12. Step changes (up/down) of the three inputs: the stack current I_{st}, the blower signal u_{blo}, and the fuel valve signal u_{valve}, are applied individually at time 400, 800, and 1200 sec, respectively, followed by the simultaneous step changes of all inputs at 1600 sec (see top three plots of Figure 6.8). The input u_{valve} has a value between

0 and 1 in these plots. Note here that, in practice, it is unlikely that an input is applied individually. Often, the blower and the valve inputs are applied simultaneously based on the changes in load current. It is therefore more critical to obtain good agreement on the responses in the case of simultaneous inputs (at 1600 sec).

The responses of the key variables are shown in Figure 6.8. In the right column is the zoom-in of the response at 1600 sec which represents the simultaneous input step increase. Various pressure and flow variables are shown in Figure 6.9. It can be seen that, despite the offset, there is a good agreement between the two models for most transient responses.

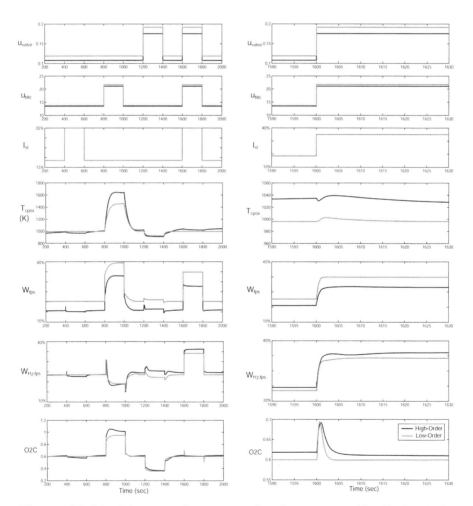

Fig. 6.8. Model validation results: inputs and performance variables. Dark = high-order model; light = low-order model

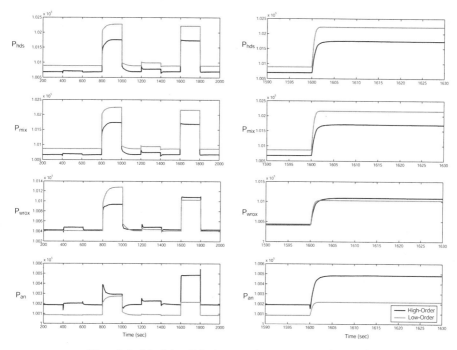

Fig. 6.9. Model validation results: pressures (Pascal)

At 400 sec where the step of stack current is applied, the low-order model does not show any transient because the stack current only affects the hydrogen consumption in the anode which has very little influence on the FPS variables. On the other hand, the high-order model shows a small transient. This transient is caused by a built-in feedforward controller in the high-order model that adjusts several flow rates based on the changes in stack current. The feedforward controller is not implemented in the low-order model.

The step of the air blower command at 800 sec raises CPOX temperature, which is a result of an increase in the O_2-to-CH_4 ratio. When O_2 to CH_4 ratio rises, the rate of the POX reaction decreases, and thus lowers the final product hydrogen, as shown in the response of $W_{H_2,fps}$. However, there is an initial increase of $W_{H_2,fps}$ right at 400 sec caused by the increase of total flow that initially has a high H_2 concentration. This behavior indicates that the FPS plant has nonminimum phase (NMP) relation from the blower command to the H_2 generation. This NMP response can also be observed when the blower command decreases, as seen from the $W_{H_2,fps}$ response at 1000 sec.

During the step increase in fuel valve command (at 1200 sec), the O_2-to-CH_4 ratio drops and results in more POX reaction, thus more hydrogen is generated ($W_{H_2,fps}$ increases). After the initial increase in $W_{H_2,fps}$, the TOX reaction drops and heat generated from the reaction is not sufficient to maintain the CPOX temperature. The drop in T_{cpox} later lowers the rate of

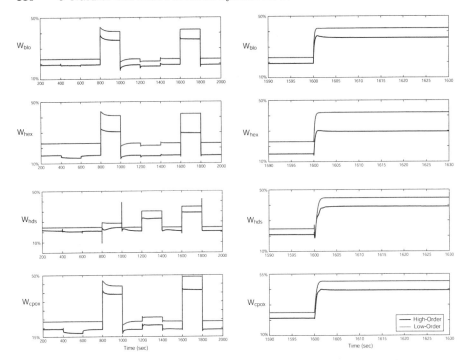

Fig. 6.10. Model validation results: flow rates (% of full flow)

CH_4 reaction (see Figure 6.5) and thus reduces the product hydrogen; that is, $W_{H_2,fps}$ decreases.

Small discrepancies can be spotted in the responses, for example, in the anode pressure. The discrepancies arise mostly from the results of the assumptions used to simplify the model. The main assumption is that the WGS and PROX reactors are combined into one volume. This results in a crude approximation of the pressure in the WGS and PROX, as can be seen in Figure 6.11.

Steady-state offsets of the model can also be reduced. For example, the CPOX temperature offset shown in Figure 6.8 might be reduced if the enthalpy terms in the CPOX temperature equation (6.48) are directly calculated from gas composition and temperature rather than from a lumped specific heat.

It can be seen that, despite the offset, there is a good agreement between the two models for most transient responses. The model is also tested at different power (current) operations and transient responses also agree well. The low-order model is therefore accurate enough for control design in Chapter 7. A more accurate model can be developed with the expense of extra complexity.

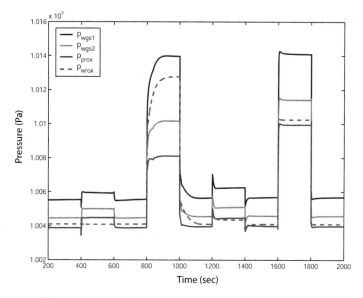

Fig. 6.11. Model validation results: WROX pressure

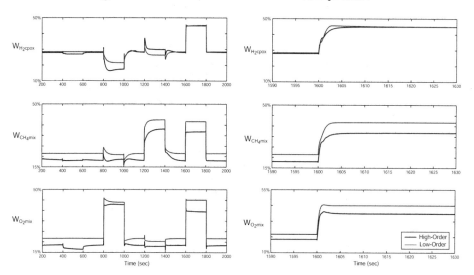

Fig. 6.12. Model validation results: CPOX composition

Table 6.3. Typical model parameters for a 200kW system [23, 32, 36, 40, 95]

Parameter	Typical Value
T_{hex}	400–500 K
V_{hex}	0.05 m^3
$W_{0,hex}$	0.04 kg/s
$\Delta p_{0,hex}$	450–500 Pa
T_{hds}	350–400 °C
V_{hds}	0.3 m^3
p_{tank}	133 kPa
$W_{0,valve}$	0.0075 kg/s
$\Delta p_{0,valve}$	3600 Pa
$W_{0,hds}$	0.0075 kg/s
$\Delta p_{0,hds}$	100–110 Pa
T_{mix}	300°C
V_{mix}	0.03 m^3
$C_{P,bed}^{cpox}$	450 J/kg·K
m_{bed}^{cpox}	2.8 kg
$W_{0,cpox}$	0.05 kg/s
$\Delta p_{0,cpox}$	3000 Pa
η_{wrox}	20–50 %
T_{wrox}	500 K
V_{wrox}	0.45 m^3
M_{wrox}	16×10^{-3} kg/mol
$T_{wgs1,in}^{des}$	400°C
$T_{wgs2,in}^{des}$	200°C
T_{wgs1}	400°C
$W_{0,wrox}$	0.06 kg/s
$\Delta p_{0,wrox}$	2000 Pa
T_{an}	65–80°C
V_{an}	0.0045 m^3
M_{an}	27.8×10^{-3} kg/mol
n	750–1000 cells
$W_{0,an}$	0.06 kg/s
$\Delta p_{0,an}$	500–600 Pa

7

Control of Natural Gas Fuel Processor

During changes in the stack current, the fuel processor needs to (i) quickly regulate the amount of hydrogen in the fuel cell stack (anode) to avoid starvation or wasted hydrogen [79] and (ii) maintain a desired temperature of the CPOX catalyst bed for high conversion efficiency [91]. Accurate control and coordination of the fuel processor reactant flows can prevent both large deviation of hydrogen concentration in the anode and large excursion of the CPOX catalyst bed temperature.

In this chapter, the control-oriented model of the natural gas fuel processing system developed in the previous chapter is used for a multivariable control analysis and design. The two main performance variables are the anode hydrogen mole fraction [105] and the CPOX catalyst bed temperature [122]. The two control actuators are the fuel (CH_4) valve command and the CPOX air blower command. The control problem is formulated in Section 7.1 and a linearized model derived in Section 7.2 is used in the control analysis and design.

Typical fuel processing systems rely on a decentralized (single-input single-output) control of the air blower command to control CPOX temperature and, separately, the fuel valve command to control the anode hydrogen concentration. In Section 7.3, an analysis using the relative gain array method confirms the appropriateness of the traditional input-output pairs for the decentralized control. The study also shows large interactions between the two loops at high frequencies and different operating conditions. These interactions can be more efficiently handled with the multivariable control studied in Section 7.5. The linear quadratic optimal control method is used to design the controller and the state estimator that achieves a significant improvement in the CPOX temperature regulation as compared to the decentralized controller. It is shown in Section 7.5.3 that the regulation of the anode H_2 mole fraction depends strongly on the speed of the fuel valve command and the improvement in the CPOX temperature regulation is due to the coordination of the two inputs.

With realistic measurements where sensor lags are significant, the performance of the multivariable controller can degrade. The analysis of observability gramian, presented in Section 7.5.4, can be used to guide the control design and measurement selections. This observability analysis can also help in assessing the relative cost-benefit ratio in adding extra sensors for the flow and pressure measurements in the system.

7.1 Control Problem Formulation

As previously discussed, the main objectives of the FPS controller are (i) to protect the stack from damage due to H_2 starvation, (ii) to protect the CPOX from overheating, and (iii) to keep overall system efficiency high, which includes high stack H_2 utilization and high FPS CH_4-to-H_2 conversion. Objectives (ii) and (iii) are related because maintaining the desired CPOX temperature during steady-state implies proper regulation of the oxygen-to-carbon ratio which corresponds to high FPS conversion efficiency.

Two performance variables that need to be regulated are the anode hydrogen mole fraction y_{H_2} (Equation (6.60)), and the CPOX temperature T_{cpox}, calculated by integrating Equation (6.21). They are chosen based on the following rationale. High T_{cpox} can cause the catalyst bed to overheat and be permanently damaged. Low T_{cpox} results in a low CH_4 reaction rate in the CPOX [122] and potential methane slip. Large deviations in y_{H_2} are undesirable. A low value of y_{H_2} means anode H_2 starvation [105, 102] which can permanently damage the fuel cell structure. On the other hand, a high value of y_{H_2} means lower hydrogen utilization which results in a waste of hydrogen.

In this control study, we assume that all CH_4 that enters the CPOX reacts without any methane slip. Note that this assumption reduces the validity of the model for large T_{cpox} deviations. The effect of the modeling error due to this assumption can degrade the performance of the model-based controller. However, achieving one of the control goals, which is the regulation of T_{cpox}, will ensure that this modeling error remains small.

The H_2 valve actuator dynamics is ignored. The stack current I_{st} is considered as an exogenous input that is measured. Because the exogenous input is measured, we consider a two degrees of freedom controller based on feedforward and feedback, as shown in Figure 7.1. The control problem is formulated using the general control configuration shown in Figure 7.2. The two control

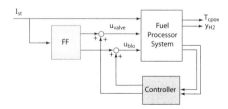

Fig. 7.1. Feedback control study

inputs u are the air blower signal u_{blo} and the fuel valve signal u_{valve}. The feedforward terms provide the valve and the blower with values that reject the steady-state effect of current to the outputs.

$$u^* = \begin{bmatrix} u_{blo}^* \\ u_{valve}^* \end{bmatrix} = f_I(I_{st}) \qquad (7.1)$$

The value of u^* is obtained by nonlinear simulation and can be implemented with a look-up table. The performance variable z includes the CPOX temper-

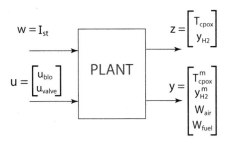

Fig. 7.2. Control problem

ature T_{cpox} and the anode exit hydrogen mole fraction $y_{H_2}^{an}$.

Several sets of measured variables are considered. The variables that can be potentially measured are the CPOX temperature T_{cpox}^m, the hydrogen mole fraction $y_{H_2}^m$, the air flow rate through the blower W_{air}, and the fuel flow rate W_{fuel}. The temperature can be measured by a thermocouple or a noncontact temperature sensor [99]. The hydrogen mole fraction can be measured with a combination of electrochemical sensors [116, 59] and model-based observers [9]. An extensive research effort is currently underway to develop fast, repeatable, and robust hydrogen sensors. The measured values T_{cpox}^m and $y_{H_2}^m$ are the values obtained from realistic sensors with measurement lag. The control objective is to reject or attenuate the response of z to the disturbance w by controlling the input u based on the measurement y.

In order to determine the fundamental limitations or issues that are related to the plant based on the actuator topology and not the sensors, we first study the control design based on the perfect measurements of the performance variables, that is, perfect measurements of T_{cpox} and $y_{H_2}^{an}$. Then, later in this chapter, we analyze the effect of realistic measurements, with sensor lag and noise, on the performance of the observer-based controller.

7.2 Analysis of FPS Linearized Models

A linear model of the FPS is obtained by linearizing the nonlinear model developed in Chapter 6. In this study, the desired steady-state is selected at stack H_2 utilization $U_{H_2} = 80\%$ [37] and CPOX oxygen-to-carbon ratio $\lambda_{O2C} = 0.6$. This condition results in the value of the CPOX temperature $T_{cpox} = 972$ K (corresponding to $\lambda_{O2C} = 0.6$), and the value of the anode hydrogen mole fraction $y_{H_2} = 8.8\%$ (corresponding to $U_{H_2} = 80\%$). The control objective is therefore to regulate T_{cpox} at 972 K and y_{H_2} at 0.088. This desired value of T_{cpox} also agrees with the value published in the literature [32]. Static feedforward terms (illustrated in Figure 7.1) are included in the linear plant so that the steady-state T_{cpox} and y_{H_2} are maintained at nominal values during changes in stack current. The linearization of the plant is given by

$$\Delta \dot{x} = A\Delta x + B_u \Delta u + B_w \Delta w$$
$$\Delta z = C_z \Delta x + D_{zu} \Delta u + D_{zw} \Delta w$$

where the state x, input u, disturbance w, and performance variables z, are

$$x = \left[T_{cpox} \; p_{H_2}^{an} \; p^{an} \; p^{hex} \; \omega_{blo} \; p^{hds} \; p_{CH_4}^{mix} \; p_{air}^{mix} \; p_{H_2}^{wrox} \; p^{wrox} \right]^T$$

$$w = I_{st} \qquad u = \left[u_{blo} \; u_{valve} \right]^T \qquad z = \left[T_{cpox} \; y_{H_2}^{an} \right]^T \qquad (7.2)$$

For simplicity, the symbol Δ, which denotes the deviation of the variables from the nominal point, is dropped for the rest of the chapter. The matrices for a 50% current (load) level are given in Table A.3. The units of states are pressure in kPa, temperature in Kelvin, and rotational speed in kRPM. The current input is in Amperes. The blower and the valve signals u_{blo} and u_{valve} vary between 0 and 100. The outputs are the CPOX temperature in Kelvin and the anode hydrogen mole fraction in percent. In the transfer function form, the plant is represented as

$$\begin{bmatrix} z \\ y \end{bmatrix} = G \begin{bmatrix} w \\ u \end{bmatrix} = \begin{bmatrix} G_{zw} & G_{zu} \\ G_{yw} & G_{yu} \end{bmatrix} \begin{bmatrix} w \\ u \end{bmatrix} \qquad (7.3)$$

The comparison of time responses between the nonlinear and the linear models in Figure A.2 shows acceptable agreements. The small offsets are the results of errors in the linearization of feedforward terms. The eigenvalues and eigenvectors of the linear system are shown in Table 7.1. It can be observed that the slow eigenvalue at –0.086 is associated with the CPOX temperature and the other slow eigenvalues at –0.358 and –1.468 are related to the hydrogen concentration in the stack anode volume and the WROX volume.

The nonlinear plant model is linearized at three different current (load) levels that correspond to 30%, 50%, and 80% of the plant power level. The linear models are referred to as 30%, 50%, and 80% systems (or models) depending on their linearization point. The Bode plots and step responses of the linear plants obtained from different system power levels are shown in Figures 7.3 and 7.4. For clarity, in these two figures, the units of current are ($\times 10$ Amp). Note first that the static feedforward controller does well in

Table 7.1. Eigenvalues and eigenvectors of FPS linear model

Eigenvalues

λ	-1.4678	-0.3579	-660.66	-157.92	-89.097	-12.171	-2.7825+0.4612i	-2.7825-0.4612i	-0.085865	-3.3333

Eigenvectors

x1	T_{cpox}	0	0	-0.002922	-0.000033	0.035363	-0.248017	-0.7546	-0.7547	0.999908	0.699066
x2	$p^{an}_{H_2}$	1	0.917378	-0.000284	0.164391	0.000331	0.138515	0.1135-0.0268i	0.1135+0.0268i	-0.006255	0.090218
x3	p^{an}	0	0	-0.000764	0.984706	-0.004846	-0.042203	-0.0024-0.0005i	-0.0024+0.0005i	0.000506	0.001705
x4	p^{hex}	0	0	0.283168	0.006251	0.367157	0.040987	0.0012-0.0072i	0.0012+0.0072i	0.001372	-0.007774
x5	ω_{blo}	0	0	0	0	0	0	0	0	0	0.000326
x6	p^{hds}	0	0	0.069280	0.000478	-0.065690	0.065140	0.0013-0.0085i	0.0013+0.0085i	0.001513	-0.009961
x7	$p^{mix}_{CH_4}$	0	0	-0.733729	-0.001551	0.710624	-0.642110	-0.4437+0.0738i	-0.4437-0.0739i	0.003010	0.492812
x8	p^{mix}_{air}	0	0	-0.613602	-0.000305	-0.595383	0.682953	-0.4449-0.0815i	-0.4449+0.0815i	-0.001495	-0.501783
x9	$p^{wrox}_{H_2}$	0	0.398016	0.003223	-0.013867	-0.006765	-0.013953	0.0991+0.0011i	0.0991-0.0011i	-0.011143	-0.091075
x10	p^{wrox}	0	0	0.011485	-0.055638	-0.009657	-0.180758	-0.011-0.0024i	-0.011+0.0024i	0.002347	0.007750

Controllability

cond(λI-A Bu)	5838.78	24130.73	12932.24	462598.60	5113.45	8426.44	5802.98	5803.98	5696.54	5800.03

Observability

cond(λI-A;Cz)	839.51	4245.69	275896.56	7375.08	17499.82	2539.83	924.46	924.46	2152.40	1000.53

rejecting the effect from I_{st} to y_{H_2} in steady-state. The H_2 recovery using feedforward is, however, relatively slow. A feedback controller is, thus, needed to speed up the system behavior and to reduce the sensitivity introduced by modeling uncertainties.

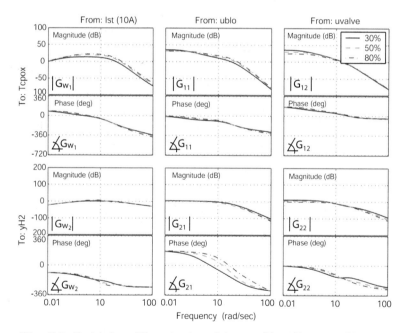

Fig. 7.3. Bode plot of linearized models at 30%, 50%, and 80% power

The response of the outputs due to step changes in the actuator signals shown in Figure 7.4 indicates a strongly coupled system. The fuel dynamics are slower than the air dynamics, primarily due to the large HDS volume.

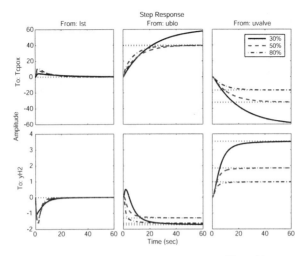

Fig. 7.4. Step responses of linearized models at 30%, 50%, and 80% power

Note that a right half-plane (RHP) zero exists in the path $u_{blo} \rightarrow y_{H_2}$; it can be easily detected from the initial inverse response of the y_{H_2} due to a step change in u_{blo}. Moreover, as can be seen in the step responses from u_{blo} to y_{H_2}, the RHP zero that causes the nonminimum phase behavior moves closer to the imaginary axis and causes a larger initial inverse response at the low power level (30%). It can also be seen in the Bode plot that, at high frequencies, the disturbance I_{st} has more effect on the H_2 mole fraction than the CPOX temperature, as compared to that at low frequencies. This is because the current has a direct impact on the amount of hydrogen used in the anode, which is coupled with the H_2 mole fraction only through the fast dynamics of the gas in the anode volume.

The characteristics of the FPS plant can vary when operating at different load levels or operating points. The distance between two system matrices is represented by the gap metric [50], which has values between zero and one. More specifically, the gap metric represents the degradation of stability margins when the first system is perturbed to become the second system [109]. A value closer to one indicates a large distance between the two systems. The gap metrics of three linear models of the plant, which are obtained by linearizing the nonlinear model at three different current (load) levels, 30%, 50%, and 80%, are presented in Table 7.2. The MATLAB® μ-Analysis Toolbox is used to calculate the gap metrics. From the large value of the gap between the 30% and 80% models, it is expected that there will be degradation of control performance when a linear controller designed for one model is used on the other. This suggests the need for gain scheduling that can be pursued in the future. The linearization of the system at the 50% power level shown in Table A.3 is used in the control study in the following sections.

Table 7.2. Gap between linearized systems

Linearization Points (power level)	Gap
30% and 50%	0.3629
50% and 80%	0.3893
30% and 80%	0.6624

7.3 Input-output Pairing

One of the most common approaches to controlling a multi-input multi-output (MIMO) system is to use a diagonal controller, which is often referred to as a decentralized controller. The decentralized control works well if the plant is close to diagonal which means that the plant can be considered as a collection of individual single-input single-output (SISO) subplants with no interactions among them. In this case, the controller for each subplant can be designed independently. If an off-diagonal element is large, then the performance of the decentralized controller may be poor.

The design of a decentralized controller involves two steps: input-output pairing and controller tuning. Interactions in the plant must be considered as we choose input-output pairs. For example, having a choice, one would drop the pairing between u_{blo} and y_{H_2} due to the RHP nonminimum phase relationship. A method used to measure the interactions and assess appropriate pairing is called the Relative Gain Array (RGA) [22]. RGA is a complex nonsingular square matrix defined as

$$RGA(G) = G \times (G^{-1})^T \tag{7.4}$$

where $\times$ denotes element by element multiplication. Each element of the RGA matrix indicates the interaction between the corresponding input-output pair. It is preferred to have a pairing that gives an RGA matrix close to an identity matrix. The useful rules for pairing are defined in [101].

1. To avoid instability caused by interactions at low frequencies one should *avoid* pairings with negative steady-state RGA elements.
2. To avoid instability caused by interactions in the crossover region one should *prefer* pairings for which the RGA matrix in this frequency range is close to identity.

The 2×2 RGA matrices of G_{zu} defined in (7.2) to (7.3) and calculated for the 50% load in Section 7.2 are given in (7.5) for different frequencies. According to the first rule, it is clear that the preferred pairing choices are the $u_{blo} \rightarrow T_{cpox}$ pair and the $u_{valve} \rightarrow y_{H_2}$ pair to avoid instability at low frequencies.

$$RGA(0 \text{ rad/s}) = \begin{bmatrix} 2.302 & -1.302 \\ -1.302 & 2.302 \end{bmatrix}$$

$$RGA(0.1 \text{ rad/s}) = \begin{bmatrix} 2.1124 - 0.36663i & -1.1124 + 0.36663i \\ -1.1124 + 0.36663i & 2.1124 - 0.36663i \end{bmatrix} \quad (7.5)$$

$$RGA(1 \text{ rad/s}) = \begin{bmatrix} 1.1726 - 0.50797i & -0.17264 + 0.50797i \\ -0.17264 + 0.50797i & 1.1726 - 0.50797i \end{bmatrix}$$

$$RGA(10 \text{ rad/s}) = \begin{bmatrix} 0.24308 - 0.0021386i & 0.75692 + 0.0021386i \\ 0.75692 + 0.0021386i & 0.24308 - 0.0021386i \end{bmatrix}$$

However, it can be seen that at high frequencies, the diagonal and off-diagonal elements are closer, which indicates more interactions. In fact, a plot of the magnitude difference between the diagonal and off-diagonal elements of the RGA matrices of the linearized systems at 30%, 50%, and 80% power in Figure 7.5 shows that interactions increase at high frequencies [81, 101]. At low power levels, the values of the off-diagonal elements of the RGA matrix are even higher than the diagonal elements ($|RGA_{11}| - |RGA_{12}| < 0$), indicating large coupling in the system. At these frequencies, we can expect poor performance from a decentralized controller.

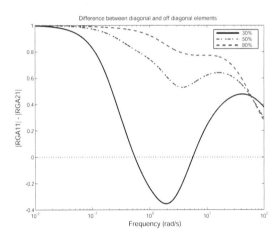

Fig. 7.5. Difference between diagonal and off-diagonal elements of the RGA matrix at different frequencies for three power setpoints

7.4 Decentralized Control

To illustrate the effect of the interactions, we designed several PI controllers for the two single-input single-output systems that correspond to the diagonal subsystem of G_{zu}, that is, $u_{blo} \rightarrow T_{cpox}$ ($G_{zu}(1,1)$) and $u_{valve} \rightarrow y_{H_2}$ ($G_{zu}(2,2)$). The PI controller is the most commonly used controller for process control. The diagram in Figure 7.6 shows the decentralized controller.

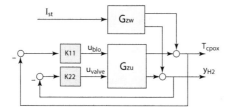

Fig. 7.6. Decentralized control

Each controller gives different closed loop characteristics, as shown in Tables 7.3 and 7.4. The closed loop Bode plots and step responses of the system

Table 7.3. Controller for $G_{zu}(1,1)$

Controller	Transfer Function	Rise Time (s)	Overshoot (%)
K11a	$0.0389\dfrac{(4.1s+1)}{s}$	3.14	11
K11b	$0.0667\dfrac{(4.3s+1)}{s}$	1.0	7
K11c	$0.0135\dfrac{(5.6s+1)}{s}$	6.58	12

Table 7.4. Controller for $G_{zu}(2,2)$

Controller	Transfer Function	Rise Time (s)	Overshoot (%)
K22a	$0.268\dfrac{(2.8s+1)}{s}$	3.95	6
K22b	$0.165\dfrac{(21s+1)}{s}$	1.33	10

with different controllers are shown in Figures 7.7 to 7.12. Three responses are shown in each figure which are open loop or feedforward response (solid), decentralized feedback response of the full plant (dashed), and the ideal decentralized control response (dotted) which is the expected response if the off-diagonal elements of G_{zu} (see (7.3)) are zero.

Relatively slow controllers (K11a and K22a) in both loops are used for the response in Figure 7.7. It can be seen that the performance of the slow decentralized controller does not deteriorate significantly when the cross-coupling interactions are introduced. Despite its robustness, the slow controller corresponds to large y_{H_2} excursions during transient. Thus, a faster controller is needed. Figure 7.8 shows the closed loop response when faster controllers (K11b and K22b) are employed in both loops. The control performance starts deteriorating due to system interactions. Moreover, because the interaction is larger for the low power (30%) system, the performance of fast decentralized control deteriorates significantly and even destabilizes the system as shown in Figure 7.9. To prevent the adverse effect of interactions, it is possible to design the two controllers to have different bandwidths. Figure 7.10 shows the response when using a slow controller in the fuel loop (K22a) and a fast controller in the air loop (K11b). It can be observed here that the recovery speed of y_{H_2} mainly depends on the speed of the fuel valve, or fuel flow, and thus the fast H_2-valve loop is necessary. Therefore, we tune the controller K22 of the $u_{valve} \rightarrow y_{H_2}$ subsystem to achieve the desired y_{H_2} response ($|y_{H_2}| < 0.08$). To get fast y_{H_2} response while avoiding the effect of the interactions, the T_{cpox}-air loop needs to be much slower or faster than the $u_{valve} \rightarrow y_{H_2}$ closed loop subsystem. Unfortunately, faster $u_{blo} \rightarrow T_{cpox}$ is not feasible due to actual magnitude constraints. Thus K11 = K11c is selected, which slows down the first subsystem loop compromising the T_{cpox} response, as shown in Figures 7.11 and 7.12. These two figures show that large time-scale separation is needed in order to use the decentralized control method. In the following sections, for comparison with other controllers, the PI controllers K11c and K22b are used.

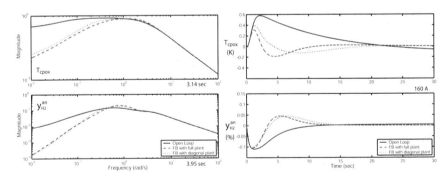

Fig. 7.7. Bode magnitude and unit step response of 50% model with controllers K11a and K22a

Note that more complex decentralized controllers can be used (PID or high-order, for example). The PI controller tuning here (Tables 7.3 and 7.4) is used only to illustrate the effect of plant interactions and difficulties in tuning the PI controllers without systematic MIMO control tools. The conclusion

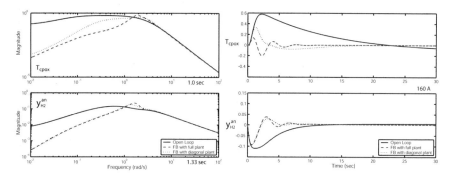

Fig. 7.8. Bode magnitude and unit step response of 50% model with controllers K11b and K22b

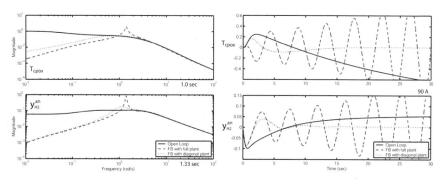

Fig. 7.9. Bode magnitude and unit step response of 30% model with controllers K11b and K22b

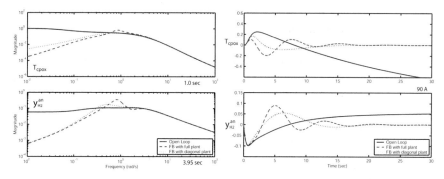

Fig. 7.10. Bode magnitude and unit step response of 30% model with controllers K11b and K22a

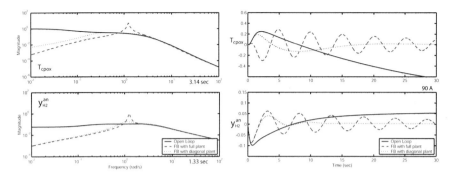

Fig. 7.11. Bode magnitude and unit step response of 30% model with controllers K11a and K22b

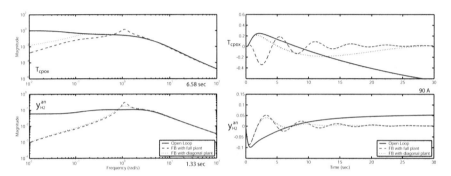

Fig. 7.12. Bode magnitude and unit step response of 30% model with controllers K11c and K22b

from this section is that the large plant interactions illustrated by Figure 7.5 must be considered in the control design.

An interesting point is that, for the decentralized PI controller, the bandwidth of the air loop needs to be smaller than the bandwidth of the fuel loop. This is the only way to achieve a bandwidth separation within the blower saturation constraints. However, if a higher-order controller is allowed, a higher closed loop bandwidth can be achieved. Indeed, as we show later in Section 7.5.3, a high-order decentralized controller using the diagonal terms of a full MIMO controller achieves a decade higher bandwidth compared to the fuel loop without saturating the blower.

7.5 Multivariable Control

We show in the previous section that interactions in the plant limit the performance of decentralized controllers. In this section, we assess the improvement that can be gained by using a multi-input multi-output (MIMO) controller and model-based control design techniques. The controller is designed using the linear quadratic (LQ) methodology. The control development consists of two steps. A full state feedback controller is designed using linear quadratic optimization for the controller gains. Because the plant states cannot be easily measured, the second step is to build a state observer or state estimator based on the measured performance variables.

7.5.1 Full-state Feedback with Integral Control

To eliminate steady-state errors, we add to the controller the integrators on the two performance variables T_{cpox} and y_{H_2}. Note that in this section, we assume that these two variables can be directly and instantaneously measured. The state equations of the integrators are

$$\frac{d}{dt}\begin{bmatrix} q_1 \\ q_2 \end{bmatrix} = \begin{bmatrix} T_{cpox}^{ref} - T_{cpox} \\ y_{H_2}^{ref} - y_{H_2} \end{bmatrix} \tag{7.6}$$

where $T_{cpox}^{ref} = 972$ K and $y_{H_2}^{ref} = 8.8\%$ are the desired values of T_{cpox} and y_{H_2}, respectively. In the linear domain, the desired deviation from the reference value is zero for all current commands. The augmented plant, which combines the original states x and the integrator states q is represented by

$$\dot{x}_a = \begin{bmatrix} \dot{x} \\ \dot{q} \end{bmatrix} = \begin{bmatrix} A & 0 \\ C_z & 0 \end{bmatrix} \begin{bmatrix} x \\ q \end{bmatrix} + \begin{bmatrix} B_u \\ D_{zu} \end{bmatrix} u = A_a x_a + B_a u \tag{7.7}$$

The controller is designed to minimize the cost function

$$J = \int_0^\infty z^T Q_z z + q^T Q_I q + u^T R u \; dt \tag{7.8}$$

where $u = [u_{blo} \; u_{valve}]^T$ and Q_z, Q_I, and R are weighting matrices on the performance variables z, integrator state q, and control input u, respectively. The cost function can be written in the linear quadratic form of the augmented states x_a as

$$J = \int_0^\infty x_a^T \begin{bmatrix} C_z^T Q_z C_z & 0 \\ 0 & Q_I \end{bmatrix} x_a + u^T R u \; dt = \int_0^\infty x_a^T Q x_a + u^T R u \; dt \tag{7.9}$$

The control law that minimizes (7.9) is in the form

$$u = -K_P(x - x_d) - K_I q = -K \begin{bmatrix} (x - x_d) \\ q \end{bmatrix} = -R^{-1} B_a^T P \begin{bmatrix} (x - x_d) \\ q \end{bmatrix} \tag{7.10}$$

where P is the solution to the Algebraic Riccati Equation (ARE)

$$P A_a + A_a^T P + Q - P B_a R^{-1} B_a^T P = 0 \tag{7.11}$$

which can be solved using MATLAB®. Variable x_d in (7.10) is a function of w and can be viewed as the desired value of the states (as a function of w)

that gives the desired value of $z = 0$. In other words, the term $K_p x_d$ is an additional feedforward term (or pre-compensator) [42, 43] that compensates for the changes in the output steady-state value due to the feedback. As a result, this additional feedforward term is a function of the feedback gain K_p. It can be shown that the term $K_p x_d$ is equivalent to the term u_p in Equation (5.17). The value of x_d can be found by simulation or by the linear plant matrices; that is,

$$x_d = \left[A^{-1} B_w \right] w \tag{7.12}$$

which results in

$$x_d^T = 10^{-3} \times \left[0 \; 0.67 \; 7.538 \; 84.148 \; 7.941 \; 79.219 \; 18.197 \; 59.761 \; 8.9 \; 35.72 \right] w \tag{7.13}$$

As it is based on the linear model, the value of x_d calculated in (7.12) will be different from the actual desired state in the nonlinear plant. The error in x_d definitely influences the steady-state error of the performance variables T_{cpox} and y_{H_2}. The integral control implemented through the augmented integrators (7.6) then becomes more critical. The fact that x_d is not accurate must be taken into account when choosing the weighting between Q_z and Q_I. Large integrator gain slows down the response, thus relatively small Q_I shows a better (faster) performance in the linear design. However, the response in nonlinear simulation with small Q_I gives poor steady-state performance because the performance is based heavily on the proportional part of the controller and therefore suffers from the error in x_d. Thus, if a more accurate value of x_d cannot be obtained, the transient performance must be compromised in order to get satisfactory steady-state performance of the controller through the integral part. Alternatively, a more accurate x_d can be obtained by numerically solving the nonlinear simulation and storing the solution in a look-up table.

Figures 7.13 and 7.14 show the closed loop responses when different weighting matrices are used in the LQ design. The effect of varying u_{valve}, while constraining the magnitude of u_{blo}, is shown in Figure 7.13. When the magnitude of u_{valve} increases, a faster response of y_{H_2} can be achieved with a small degradation of the T_{cpox} response. On the other hand, Figure 7.14 shows a large tradeoff between two performance variables when the magnitude of u_{blo} is varied while maintaining the magnitude of u_{valve}. It can be seen that a small improvement in y_{H_2} can be obtained using u_{blo}. However, there is a large degradation in the T_{cpox} response because large u_{blo} is needed to improve y_{H_2} (due to the nonminimum phase relation), and thus significantly affects T_{cpox}. These two figures imply that the improvement in H_2 starvation (y_{H_2}) mainly depends on the speed and magnitude of the valve command u_{valve}. The blower command u_{blo} has little impact on y_{H_2}, but if well coordinated with the valve command, it can provide a large improvement in T_{cpox} regulation.

The final design of the controller generates the response shown in Figure 7.15. The controller gains are obtained by using the weighting matrices

$$Q_z = \begin{bmatrix} 80 & 0 \\ 0 & 1100 \end{bmatrix} \quad Q_I = \begin{bmatrix} 150 & 0 \\ 0 & 100 \end{bmatrix} \quad R = \begin{bmatrix} 100 & 0 \\ 0 & 120 \end{bmatrix} \tag{7.14}$$

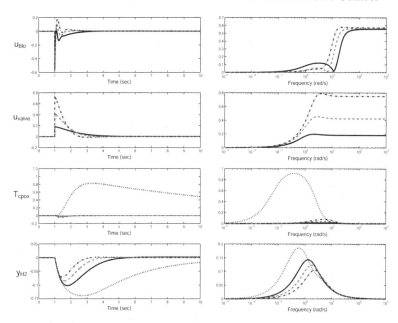

Fig. 7.13. Tradeoff between two performance variables T_{cpox} and y_{H_2} with respect to magnitude of u_{valve}. The plots are created by selecting different weighting matrices in the LQ design

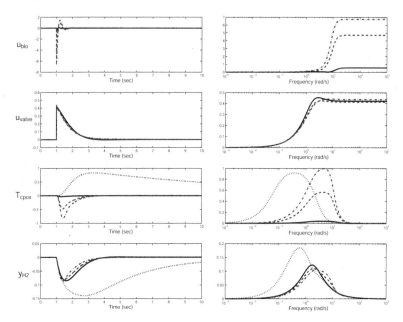

Fig. 7.14. Tradeoff between two performance variables T_{cpox} and y_{H_2} with respect to magnitude of u_{blo}. The plots are created by selecting different weighting matrices in the LQ design

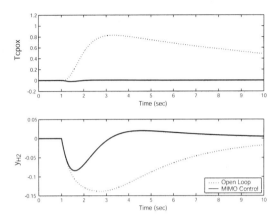

Fig. 7.15. Response of the FPS linear model with the controller from LQ design

which results in the gains

$$K_P = \begin{bmatrix} 1.405 & 0.182 & 0.029 & 1.066 & 39.04 & -6.611 & -0.705 & 0.604 & 0.767 & 0.939 \\ -0.130 & 1.150 & -0.132 & -0.244 & -8.422 & 6.111 & 0.618 & -0.139 & 3.787 & 0.127 \end{bmatrix}$$

$$K_I = \begin{bmatrix} -1.207 & -0.169 \\ 0.189 & -0.9 \end{bmatrix} \quad (7.15)$$

The closed loop eigenvalues are given in Equation (A.5).

There are a few slow closed loop eigenvalues which are the result of the weakly controllable mode associated with the plant eigenvalue $\lambda_2 = -0.3579$, as suggested by the large condition number of $[\lambda_2 I - A\; B_u]$ in Table 7.1.

A comparison of the decentralized controller, explained in Section 7.4, and the full-state feedback controller in nonlinear simulation is shown in Figure 7.16. The significant improvements in both T_{cpox} and y_{H_2} regulation when using the MIMO controller are the result of considering system interactions via the model-based state feedback design. To be able to implement the MIMO controller, in the next section, the full-state feedback controller is converted into output feedback using available measurements.

7.5.2 State Estimator

The estimate of the plant state $\hat{x}$ can be determined using the dynamic model of the plant together with the available performance measurements. It is assumed that perfect measurements of T_{cpox} and y_{H_2} are available. The observer state equations are

$$\dot{\hat{x}} = A\hat{x} + B_u u + B_w w + L(z - \hat{z})$$
$$\hat{z} = C_z \hat{x} + D_{zu} u + D_{zw} w \quad (7.16)$$

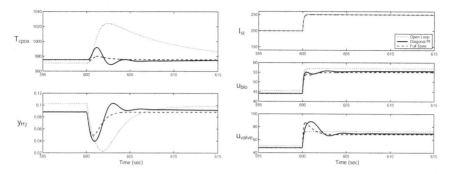

Fig. 7.16. Comparison of decentralized control and full-state feedback control in nonlinear simulation

where $\hat{x}$ is the estimator state vector and L is the estimator gain. The observability gramian Q_{obs}, that is, the solution of

$$A^T Q_{obs} + Q_{obs} A = -C_z^T C_z \qquad (7.17)$$

has full rank but the condition number of the observability gramian is high, indicating rank deficiency (*i.e.*, weak observability). Sometimes, this result arises because of poor selection of units of the model states (scaling). Thus, to better evaluate system observability, we normalize the condition number of the observability gramian (c_{obs}^N) by the value when all the states are measured, $C_x = I$.

$$c_{obs}^N = \frac{\text{cond}\left(Q_{obs,\{C=C_z\}}\right)}{\text{cond}\left(Q_{obs,\{C=I\}}\right)} = 2 \times 10^5 \qquad (7.18)$$

A large normalized observability gramian implies that the pair (A, C_z) is weakly observable.

The observer gain L in (7.16) is used to place the observer eigenvalue at the desired points, which is normally at least twice as fast as the dominant closed loop eigenvalues. Because the plant has several fast eigenvalues (Table 7.1) that do not need to be moved using output feedback, a reduced-order observer can be used to simplify the observer. This can be done by first transforming the system matrices into the modal canonical form [26] $x_1 = Tx$ such that the new system matrices are

$$A_1 = TAT^{-1} = \begin{bmatrix} \lambda_1 & & 0 \\ & \ddots & \\ 0 & & \lambda_{10} \end{bmatrix} \qquad (7.19)$$

$$C_1 = C_z T^{-1} \quad B_1 = [\, B_{1w} \; B_{1u} \,] = T\,[\, B_w \; B_u \,] \quad D_1 = [\, D_{zw} \; D_{zu} \,] \quad (7.20)$$

Note the special structure of the matrix A_1 which has eigenvalues on the diagonal. The system matrices in the new coordinates are shown in Table A.4. The matrices are then partitioned into

$$\begin{bmatrix} A_{1\bar{o}} & 0 \\ 0 & A_{1o} \end{bmatrix} \qquad \begin{bmatrix} B_{1\bar{o}} \\ B_{1o} \end{bmatrix} \qquad \begin{bmatrix} C_{2\bar{o}} & C_{2o} \end{bmatrix} \tag{7.21}$$

where

$$A_{1o} = \begin{bmatrix} -3.333 & 0 & 0 & 0 & 0 & 0 \\ 0 & -2.782 & 0.4612 & 0 & 0 & 0 \\ 0 & -0.4612 & -2.782 & 0 & 0 & 0 \\ 0 & 0 & 0 & -1.468 & 0 & 0 \\ 0 & 0 & 0 & 0 & -0.358 & 0 \\ 0 & 0 & 0 & 0 & 0 & -0.086 \end{bmatrix} \tag{7.22}$$

which contains the slow eigenvalues of the plant. The reduced-order observer gain L_1 is then designed for the set $A_{1x} = A_{1o} + \alpha I$, B_{1o}, and C_{1o}. The modification of A_{1o} to A_{1x} follows the method described in [7] to guide the observer pole placement for fast response as prescribed by the constant α. Using the Kalman filter method, the observer gain L_1 is determined by solving the optimal quadratic problem

$$L_1 := SC_{1o}^T W_y^{-1} \tag{7.23}$$

$$0 = SA_{1o}^T + A_{1o}S + V_x + SC_{1o}^T W_y^{-1} C_{1o}S \tag{7.24}$$

The weighting matrices V and W represent the process noise and measurement noise, respectively. The weighting matrices chosen are

$$V_x = 100 \operatorname{diag} \begin{bmatrix} 10 & 200 & 200 & 20 & 50 & 80 \end{bmatrix} + B_{1o} B_{1o}^T \tag{7.25}$$

$$W_y = 1 \times 10^{-6} \operatorname{diag} \begin{bmatrix} 0.1 & 0.01 \end{bmatrix} \tag{7.26}$$

The reduced-order observer gain L_1 is then transformed to the original coordinate L,

$$L = T^{-1} \begin{bmatrix} 0_{4 \times 2} \\ L_1 \end{bmatrix} \tag{7.27}$$

which results in the gain given in Equation (A.6). Figure 7.17 shows the response of the observer error $(x - \hat{x})$ in a linear simulation. The initial errors of all states are set at 1% of the maximum deviation from the nominal point. It can be seen that most estimation errors disappear within 1 sec. The nonlinear simulation of the system with the decentralized PI feedback and with the output observer-based feedback (MIMO) is shown in Figure 7.18. The output feedback gives satisfactory performance in both y_{H_2} and T_{cpox} regulations.

7.5.3 Insight Gained by the Multivariable Design

The combination of the state feedback control (7.10) and the state observer (7.16) results in a model-based multivariable output-feedback controller. The state space representation of the controller can be written as

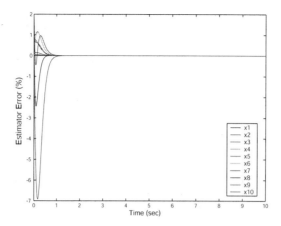

Fig. 7.17. Estimator error for 1% initial error

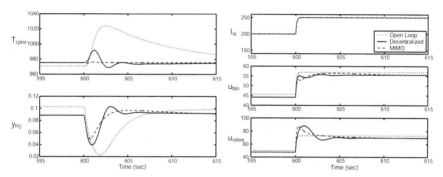

Fig. 7.18. Comparison of decentralized PI controller and observer feedback in non-linear simulation

$$\begin{bmatrix} \dot{\hat{x}} \\ \dot{q} \end{bmatrix} = \begin{bmatrix} A - B_u K - L C_z & -B_u K_I \\ 0 & 0 \end{bmatrix} \begin{bmatrix} \hat{x} \\ q \end{bmatrix} + \begin{bmatrix} B_w + B_u K_P \alpha_d & L \\ 0 & -I \end{bmatrix} \begin{bmatrix} w \\ z \end{bmatrix}$$

$$u = \begin{bmatrix} -K_p & -K_I \end{bmatrix} \begin{bmatrix} \hat{x} \\ q \end{bmatrix} + \begin{bmatrix} K_p \alpha_d & 0 \end{bmatrix} \begin{bmatrix} w \\ z \end{bmatrix} \tag{7.28}$$

where the output of the controller is the plant input $u = [\, u_{blo} \; u_{valve} \,]^T$. Variable α_d is equal to the coefficient of x_d in Equation (7.13). Note that Equation (7.28) is formed taking into account that $D_{zu} = D_{zw} = 0$. In transfer function form, the controller can be written as

$$u = C_{uw} w + C_{uz} z = \begin{bmatrix} C_{w1} \\ C_{w2} \end{bmatrix} w + \begin{bmatrix} C_{11} & C_{12} \\ C_{21} & C_{22} \end{bmatrix} z \tag{7.29}$$

The Bode plot of each element of the controller C is shown in Figure 7.19. The C_{uw} term is an additional dynamic feedforward that is superimposed on the original static feedforward u^* in (7.1). The term C_{uz} is the feedback part of the controller.

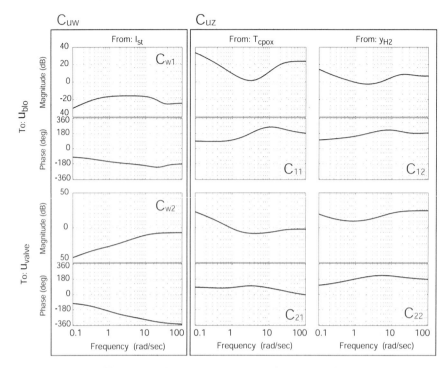

Fig. 7.19. Frequency response of the controller

In an effort to simplify the feedback controller for gain scheduling and implementation purposes, we investigate which cross-coupling term of the feedback contributes to the performance improvement by the MIMO controller. By zeroing out the cross-coupling term and plotting the closed loop frequency and time responses in Figures 7.20 and 7.21, we can see that the performance of the full controller is maintained when $C_{21} = 0$ (triangular MIMO controller). However, the performance degrades when $C_{12} = 0$ (diagonal MIMO controller). Thus it is clear that the C_{12} term is the critical cross-coupling term that provides the MIMO control improvement. This analysis gives a different result, however, if the blower bandwidth is allowed to be higher, for example, by using a more powerful and faster blower, as shown in Figures 7.22 and 7.23. These plots are generated by lowering the LQ weight on u_{blo} in the state feedback design; that is, R(1,1) = 1 in (7.14). There is more actuator activity (high-bandwidth controller) of u_{blo}, as shown in Figure 7.22, and the diagonal controller ($C_{12} = C_{21} = 0$) performs similarly to the full multivariable controller.

The importance of C_{12} is interpreted as follows. Following Figure 7.24, the current disturbance I_{st} affects y_{H_2} more than T_{cpox} during fast transient as can be seen by the large high-frequency magnitude of the transfer function from I_{st} to y_{H_2} (Figure 7.25) for the plant with feedforward control: $\tilde{G}_w =$

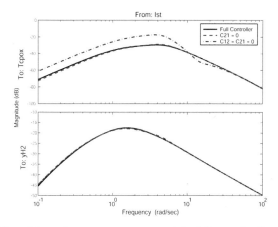

Fig. 7.20. Closed loop frequency response with different feedback controllers

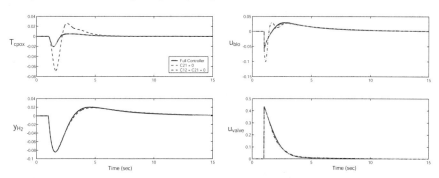

Fig. 7.21. Closed loop time response for analysis of elements in the feedback controllers

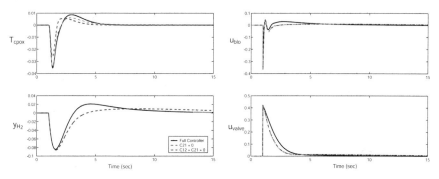

Fig. 7.22. Closed loop time response for analysis of elements in the feedback controllers assuming high-bandwidth blower

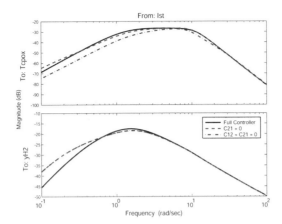

Fig. 7.23. Closed loop frequency response for analysis of elements in the feedback controllers with high-bandwidth blower

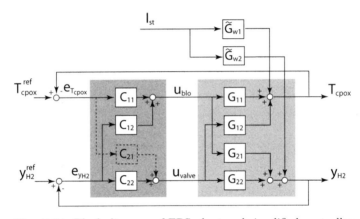

Fig. 7.24. Block diagram of FPS plant and simplified controller

$G_w + GC_w$. The valve signal u_{valve} tries to reject the effect of $\tilde{G}_{w2}I_{st}$ to y_{H_2} (see Figure 7.24) by using the feedback terms C_{22} through G_{22}. The blower signal, on the other hand, cannot reject the effect of $\tilde{G}_{w2}I_{st}$ to y_{H_2} through $G_{21}C_{12}$ because of the nonminimum phase zero of G_{21} ($z_{NMP} = 3.07$). This can be verified by the equality of two frequency plots close to the nonminimum phase frequency:

$$\frac{\tilde{G}_{w2}I_{st}}{1 + G_{21}C_{12} + G_{22}C_{22}} \simeq \frac{\tilde{G}_{w2}I_{st}}{1 + G_{22}C_{22}}$$

Indeed, Figure 7.19 shows that the magnitude of C_{12} is low at frequencies close to that of the NMP zero. Meanwhile, the valve that tries hard to reject the $\tilde{G}_{w2}I_{st}$ to y_{H_2} causes disturbances to T_{cpox} through the plant G_{12} interaction. The controller cross-coupling term C_{12} is thus needed to compensate for the

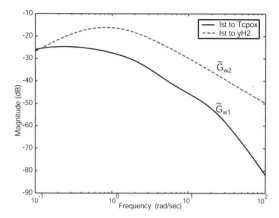

Fig. 7.25. Frequency magnitude plot of the plant with dynamic feedforward part of the controller $\tilde{G}_{zw}$ in (7.5.3)

effect of u_{valve} to T_{cpox} by partially canceling $G_{12}C_{22}$ with $G_{11}C_{12}$ at certain frequencies.

$$T_{cpox} = \frac{G_{12}C_{22} + G_{11}C_{12}}{1 + G_{11}C_{11}}e_{y_{H_2}} \approx 0 \Rightarrow C_{12} \approx -G_{11}^{-1}G_{12}C_{22}$$

Note that this partial cancellation involves the plant elements G_{11} and G_{12} that do not change significantly for different power levels, as compared to G_{21} (see Figure 7.3). Thus the benefit of the controller cross-coupling term C_{12} is maintained in the full range of operating power. If the air loop has high bandwidth, the $G_{11}C_{11}$ term can reject the disturbance by itself and, then, controller C_{12} is not needed to cancel the interaction from the valve to T_{cpox}.

Figure 7.19 also verifies that C_{21} does not contribute to the overall MIMO controller. The magnitude of C_{21} is, in fact, smaller than other feedback terms. At high frequencies where the effect of $\tilde{G}_{w2}I_{st}$ to y_{H_2} is large, the term C_{21} is not used to help regulating y_{H_2} because the deviation in y_{H_2} is not reflected in the T_{cpox} measurement ($\tilde{G}_{w1}I_{st}$ is small). At low frequencies where I_{st} affects T_{cpox}, C_{21} may be used to help reduce T_{cpox} error but will cause disturbance to the well-behaved fuel loop, thus C_{21} is also insignificant at low frequencies.

By comparing the response of the decentralized PI controller in Figure 7.18 and that of the diagonal MIMO controller in Figure 7.21, we can see that the diagonal controller derived from the MIMO controller outperforms the decentralized PI controller. This is achieved as shown in Figure 7.26 because of the higher closed loop bandwidth of the air loop when compared with that of the PI-based controller. As seen in Figure 7.19, the high-order C_{11} term achieves high bandwidth without having high gains and thus avoids blower saturation. High bandwidth cannot be achieved using a PI controller. Indeed, Figure 7.19 verifies that the gain of C_{11} is low at the frequencies where loop interaction is large (see Figure 7.5 for the loop interactions).

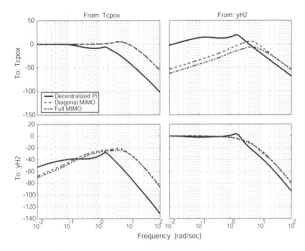

Fig. 7.26. Frequency (magnitude in dB) response from reference signal of closed loop system with MIMO controller

In summary, the MIMO controller achieves a superior performance in comparison with the decentralized PI controller due to certain factors. First, the MIMO controller achieves a high bandwidth on the air loop without saturating the actuator. This is only feasible with high-order controllers. In hindsight of the success of the C_{11} term of the MIMO controller, one can design a PID or a PI + lead-lag controller that reproduces similar gain and phase to be used in the decentralized controller.

Second, the MIMO controller achieves better coordination between the two actuators by utilizing a cross-coupling term. The cross-coupling term acts in a "feedforward" sense and changes the blower command based on how the fuel valve behaves. This partially cancels the interaction between the fuel valve to the air loop. This partial cancellation, luckily, involves plant elements that do not change significantly for different power levels. Thus, without having explicitly designed for robustness, the MIMO controller maintains its performance at all power levels.

7.5.4 Effect of Measurements

In practice, the CPOX temperature measurement and anode hydrogen mole fraction cannot be instantaneously measured. The temperature and hydrogen sensors are normally slow, with time constants of approximately 40 sec and 10 sec [108], respectively. In this section, we show that the lag in the measurements can potentially degrade the estimator performance, and thus the feedback bandwidth must be detuned in favor of robustness. For fast response, the system has to rely more on feedforward control of the fuel valve and the blower command based on the measured exogenous input I_{st}. The feedforward controller, in turn, depends on the actuator dynamics and reliability. A common method to speed up and robustify actuator performance is a cascade configuration of a 2DOF controller for each actuator based on measurement of the air flow rate W_{air} and the fuel flow rate W_{fuel} (both in g/sec). The cascade controller architecture in Figure 7.27 shows a decentralized version of the cascade controller. The decentralized version, which is very popular in industrial settings [86], uses feedforward and PI controllers to achieve the desired fuel and air flows.

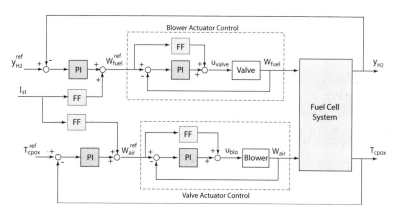

Fig. 7.27. Typical FPS control configuration

Two additional dynamic states are added to the plant when we consider the dynamics of the sensors. Two sensor-state equations that are augmented to the plant (Equation (7.2)) are

$$\begin{bmatrix} \dot{s}_T \\ \dot{s}_H \end{bmatrix} = \begin{bmatrix} -0.025 & 0 \\ 0 & -0.1 \end{bmatrix} \begin{bmatrix} s_T \\ s_H \end{bmatrix} + \begin{bmatrix} 0.025 & 0 \\ 0 & 0.1 \end{bmatrix} \begin{bmatrix} T_{cpox} \\ y_{H_2} \end{bmatrix} \quad (7.30)$$

where s_T is the CPOX temperature sensor state and s_H is the hydrogen sensor state. The order of the plant and sensor dynamics is 12 (10 for the plant and 2 for the sensors).

$$x_s = \begin{bmatrix} x^T & s_T & s_H \end{bmatrix}^T \quad (7.31)$$

where x is from Equation (7.2). The measurements are

$$y = \begin{bmatrix} T^m_{cpox} & y^m_{H_2} & W_{air} & W_{fuel} \end{bmatrix}^T = C_y x_1 + D_{yu} u + D_{yw} w \quad (7.32)$$

where T^m_{cpox} and $y^m_{H_2}$ are the measured values of T_{cpox} and y_{H_2}, respectively. The matrices in (7.32) are shown in Equation (A.7).

The observability gramian Q_{obs}, that is the solution of

$$A^T Q_{obs} + Q_{obs} A = -C_y^T C_y \qquad (7.33)$$

is used to determine the degree of system observability for a set of measurements. If the gramian has full rank, the system is observable. However, a high condition number of the observability gramian indicates weak observability. Sometimes, this result arises because of poor selection of units of the model states (scaling). Thus, to better evaluate system observability, we normalize the condition number of the observability gramian (c_{obs}^N) by the value when all the states are measured, $y = x$ or $C_y = I$. For example, the normalized observability gramian when the two performance variables are measured is:

$$c_{obs}^N = \frac{\text{cond}\left(Q_{obs}, \{y=[T_{cpox}, y_{H_2}]\}\right)}{\text{cond}\left(Q_{obs}, \{y=x\}\right)} = 2 \times 10^5 \qquad (7.34)$$

A large normalized observability gramian implies that the system with perfect measurements of T_{cpox} and y_{H_2} is weakly observable.

Each set of measurements provides a different degree of observability as can be seen by comparing the normalized condition number of the observability gramian in Table 7.5. The normalized observability gramian for slow temperature and hydrogen sensors is calculated to be 1.3×10^{10}. The lag in the measurements can potentially degrade the estimator performance, and thus the feedback bandwidth must be detuned in favor of robustness.

Adding the fuel and air flow measurements lowers the observability condition number to a value lower than the one obtained with perfect measurement of T_{cpox} and y_{H_2}. We thus expect a better estimation performance. The estimation performance is expected to be even better if additional measurements such as mixer pressure are available, as shown in Table 7.5. More work is needed to define the critical measurements that will be beneficial for the observer-based controller.

Table 7.5. Normalized condition number of observability gramian

Measurements	Condition Number
T_{cpox}, y_{H_2}	2×10^5
$T_{cpox}^m, y_{H_2}^m$	1.3×10^{10}
$T_{cpox}^m, y_{H_2}^m, W_{air}, W_{fuel}$	3672.7
$T_{cpox}^m, y_{H_2}^m, W_{air}, W_{fuel}, p^{mix}$	1928.8

8

Closing Remarks

A satisfactory transient behavior is one of the critical requirements of the fuel cell system for both automotive and residential applications. A well-designed control system is needed in order to provide fast and consistent transient behavior of the fuel cell system. The overall system for a typical polymer electrolyte membrane fuel cell (PEM-FC) consists of four main subsystems, namely, reactant supply, heat and temperature, water management, and power management subsystems. Additional complexities arise for the system with a hydrogen fuel processor that converts carbon-based fuel into hydrogen. Interactions among the subsystems lead to complex control problems.

Control problems related to the PEM fuel cell system are presented in this book. The first problem is the control of the cathode oxygen reactant for a high-pressure direct hydrogen fuel cell system (FCS). The control goal is to effectively regulate the oxygen concentration in the cathode by replenishing quickly and accurately the oxygen depleted during power generation. The second problem is the multi-input multi-output control of a low-pressure partial oxidation based natural gas fuel processor system (FPS). The control objectives are to regulate both catalytic partial oxidation (CPOX) temperature and anode hydrogen concentration. System dynamic analysis and control design are carried out using a model-based linear control approach.

A control-oriented nonlinear dynamic model suitable for each control problem is developed from physics-based principles. Not only are they easily scalable and expandable, but the system-level dynamic models built from physics-based component models are also very useful for understanding the subsystem interactions and designing model-based controllers. The models of the FCS and FPS are developed using physical principles such as chemical reaction, electrochemistry, thermodynamics, mechanics, and lumped parameter fluid dynamic principles. The transient behavior captured in the model includes flow characteristics, inertia dynamics, manifold filling dynamics, time-evolving reactant pressure or mole fraction, membrane humidity, and the relevant CPOX converter temperature.

8.1 Fuel Cell Stack System

The stack voltage is calculated based on time-varying load current, cell temperature, air pressure, oxygen and hydrogen partial pressure, and membrane humidity. The fuel cell voltage is determined using a polarization curve based on the reversible cell voltage, activation losses, ohmic losses, and concentration losses. Flow equations, mass conservation, and electrochemical relations are used to calculate changes in partial pressures and the humidity of the gas in the fuel cell stack flow channels. The FCS model contains nine states that capture the dynamics evolution of critical system variables.

In this study, we focus on the control of the cathode oxygen supply. For this purpose, a proportional controller for the hydrogen flow and an ideal humidifier are used in the FCS model. The hydrogen flow control ensures a minimum pressure difference between the anode and the cathode channels, and the humidifier ensures a fixed humidity of the air entering the stack. Furthermore, perfect conditions of temperature and humidity are assumed at several places in the model, for example, the temperature and humidity of the inlet air, of the inlet hydrogen, and of the cell membrane. The control input is the compressor command. The performance variable is the oxygen excess ratio, which is defined as the ratio between oxygen supplied to the cathode and oxygen used in the reaction.

The steady-state optimal value of the oxygen excess ratio is obtained from analysis of the nonlinear FCS model. Operating the system at the optimal value ensures that the maximum net power is achieved for a specific current load. The identified optimal value agrees with the fuel cell specification and its desired operating point given in the literature, thus indirectly validating the accuracy of the model.

Features and properties of different control configurations such as dynamic feedforward, observer feedback, and proportional plus integral are presented. The advantages and disadvantages, such as simplicity and robustness, of each configuration are explained. Depending on the characteristics of the fuel cell system and the system model, such as source of unknown disturbance, degree of parameter variations, and/or model accuracy, a control engineer can select the most suitable control configuration.

Control performance limitations due to sensor availability are also illustrated. The performance variable, that is, the oxygen excess ratio λ_{O_2}, itself, cannot be measured. The compressor flow rate, which is located upstream from the λ_{O_2} location is used. The fact that W_{cp} is measured instead of λ_{O_2} limits the uses of integral control. The two main reasons are as follows: the reference value needs to be calculated from a known atmospheric condition, which in reality varies; and a large integral gain cannot be used as it enforces a fast compressor response to the setpoint upstream from the manifold filling volume, and thus slows down the λ_{O_2} response.

Using the stack voltage measurement as one of the feedback signals to the controller increases the system observability. Voltage is currently used for monitoring, diagnostic, and emergency shut-down procedures. The observability analysis presented suggests that the voltage should be used in the feedback, especially for estimation purposes.

There is a tradeoff between fast regulation of the oxygen excess ratio and fast delivery of the desired net power during transient operations. The conflict arises from the fact that the compressor is using part of the stack power to accelerate. The tradeoff is shown to be associated with frequencies between 0.11 to 3.2 Hz for the FCS in this study. One way to resolve this conflict is to augment the FCS with an auxiliary battery or an ultracapacitor that can drive the auxiliary devices or can potentially filter current demand to frequencies lower than 0.11 Hz.

8.2 Natural Gas Fuel Processor System

A low-order (10 states) model of the FPS is developed with a focus on the dynamic behavior associated with the flow and the pressures in the FPS and the temperature of the CPOX. The effects of both the CPOX temperature and O_2-to-CH_4 ratio on the CPOX reaction are included. The FPS model is parameterized and validated with a high-order detailed fuel cell system model. The model allows us to analyze the FPS control problem. Specifically, the FPS two-input two-output (TITO) control problem has the air blower and the fuel valve as inputs and the CPOX temperature and the anode hydrogen mole fraction as performance variables.

We show that tuning two PI controllers for the air and the fuel loops is difficult. Moreover, the closed loop performance is adversely affected by the intrinsic interaction between the two loops. One way to prevent the performance degradation is to have bandwidth separation between the two control loops. This introduces a compromise of the air–temperature closed loop response in favor of the fuel–hydrogen loop.

On the other hand, a model-based high-order controller designed using linear multivariable methodologies, LQR-LQG in our case, can achieve very good response for a wide range of operating conditions. Our analysis shows that the multivariable controller can be simplified to a lower triangular controller where the blower command depends on both errors in T_{cpox} and y_{H_2} (or, equivalently, the fuel valve). If the multivariable controller is further simplified to a diagonal controller (no cross-coupling between control inputs and errors in the performance variables), the closed loop performance degrades with respect to the full multivariable controller but it still outperforms the two PI-based closed loop performance.

Additional measurements are needed if the MIMO controller is to be implemented with realistic sensors that have slow dynamics. The observability analysis can help in assessing the relative cost-benefit ratio for adding extra sensors in the system.

8.3 Future Study

Modeling: A lot of work remains to be done in the fuel cell modeling area. Even though the FCS model behavior agrees with several experimental results published in the literature, the model and its parameters have not yet been verified with experimental data from an actual fuel cell system. An extensive identification effort is needed to increase the model fidelity. Each component model, such as the compressor, the blower, or the manifolds, can be parameterized and validated separately. A considerable amount of validation is also needed for the stack model, especially for the parameters related to the humidity, for example, those used to calculate the water transfer rate across the membrane and those used to calculate the effects of membrane humidity on the cell voltage. Some of these parameters can be determined with experiments on a single cell or a single membrane. However, experimental procedures for the identification of the stack-effective membrane humidity parameters are not known. Other parameters, such as the orifice constants, can be easily obtained from a stack-level experiment.

The fuel cell voltage also depends on the liquid water residing in the fuel cell stack. Accumulation of liquid water in the fuel cell, also know as a "flooding phenomenon" needs to be included in the model. Several publications [14, 15, 76] suggest that cathode water flooding reduces the fuel cell performance because it decreases the porosity of the electrode, which affects the diffusion ability of the gas. Temperature changes also have a significant impact on the humidity of the gases and the membrane. The temperature effects need to be taken into account in the model by either developing dynamic models of the gas and stack temperatures or, if the temperature dynamics is considered slow, analyzing the system behavior at different temperature setpoints.

A substantial amount of information is lost when using the lumped-parameter models. Important fuel cell variables such as partial pressures and temperature are, in fact, spatially distributed along the flow channel. As a result, the current density is not uniformly distributed over the fuel cell area. The effect of spatial variation needs to be included in the model especially if the model is to be used for estimation or diagnostic purposes. Flow patterns must also be incorporated to improve estimation accuracy.

FC stack control: Several interesting control problems can be addressed using the existing model. In a typical fuel cell operation, extra hydrogen is supplied to the stack in order to avoid hydrogen starvation at the end of the anode channel and excessive flooding. Thus, there is always unused hydrogen leaving the stack. To make use of the remaining hydrogen, the anode recirculation, in which the exit hydrogen flow is rerouted to the anode inlet flow, is implemented similarly to that in the P2000 system [1]. This recirculation improves the steady-state hydrogen utilization, and thus system efficiency. However, the recirculation may magnify the difficulties in controlling the anode hydrogen concentration during transient. This is due to the additional volume associated with the re-circulation. It might be interesting to analyze the dynamic behavior of the anode recirculation process and the tradeoffs associated with the steady-state and transient behavior.

The problem of online finding the steady-state optimal oxygen excess ratio will be useful in enhancing the efficiency of the fuel cell system. The optimal value of the excess ratio varies with different operating loads and may change depending on system age and environmental conditions. Extremum-seeking or other maximum-finding techniques can be used to search online for the optimum excess ratio levels. Apart from supplying enough oxygen reactants, excess air is also required to manage flooding. These multiobjective considerations result in a challenging control problem especially in low-pressure fuel cell systems that utilize blowers instead of high-speed compressors. The coupling between the heat and the water management system requires special attention and is expected to be crucial for low-pressure FCS.

FPS control: The main assumptions of the FPS model used in the control study are: (1) full conversion of CH_4, and (2) a combined lumped volume of the water gas shift (WGS) and preferential oxidation (PROX) reactors. It is desirable to eliminate these FPS assumptions in order to improve the fidelity of the model. However, if the first assumption is relaxed, the FPS model will become highly nonlinear with respect to the O_2-to-CH_4 ratio and the CPOX temperature. Thus, nonlinear techniques are needed to analyze and design the controller for the system. The assumptions on the WGS and the PROX volume limit the range of predictions for the anode variables. Separate dynamic models of the two reactors can be added in the current FPS model offering higher fidelity at the expense of model complexity. The need for precise control of carbon monoxide in the anode inlet gas requires coordination of additional actuator inputs such as WGS water flow and PROX air flow. The problem becomes even more challenging if the supply of PROX air flow is shared with that of the CPOX air flow, which is desired in commercial applications due to space and cost constraints.

The fuel processing approach discussed in this book is based on catalytic partial oxidation of natural gas fuel for hydrogen generation. Other types of fuels, such as gasoline and methanol, are also considered as potential fuels for fuel cells. There are also different types of hydrogen conversion processes such as steam reforming and auto-thermal reforming. The control task is then augmented by the fuel vaporization and steam generation which are slow and heat-intensive processes.

Last but not least, higher overall system (FPS and FCS) efficiency can be achieved by using the anode outlet gas in a catalytic burner to heat the inlet CPOX fuel and air flow, and vaporize the WGS water flow. Combined optimization of heat and power (CHP) in the fuel cell and the catalytic burner might allow lower FCS utilization levels, which can alleviate the transient requirements for hydrogen generation. However, tight integration of the fuel cell, the catalytic burner, and all the heat exchangers in the FPS creates thermal feedback loops and challenging control problems. Similar concurrent control-optimization problems arise in Solid Oxide Fuel Cells (SOFC) when they are combined with gas turbines for stationary power and heavy transportation industry (rail and marine) applications.

A

Miscellaneous Equations, Tables, and Figures

A.1 FCS Air Flow Control Design

This section presents the tables, figures, and equations related to the FCS control design presented in Chapter 5. Tables A.1 and A.2 show the system matrices obtained by linearizing the FCS nonlinear model with and without static feedforward. Figure A.1 shows that the responses from nonlinear and linearized models agree very well.

Table A.1. Linearization results without static feedforward

A

-6.30908	0	-10.9544	0	83.74458	0	0	24.05866
0	-161.083	0	0	51.52923	0	-18.0261	0
-18.7858	0	-46.3136	0	275.6592	0	0	158.3741
0	0	0	-17.3506	193.9373	0	0	0
1.299576	0	2.969317	0.3977	-38.7024	0.105748	0	0
16.64244	0	38.02522	5.066579	-479.384	0	0	0
0	-450.386	0	0	142.2084	0	-80.9472	0
2.02257	0	4.621237	0	0	0	0	-51.2108

B_u

0
0
0
3.946683
0
0
0
0

B_w

-0.03159
-0.00398
0
0
0
0
-0.05242
0

C_z

-2.48373	-1.9773	0.109013	-0.21897	0	0	0	0
-0.63477	0	-1.45035	0	13.84308	0	0	0

D_{zu}

0.169141
0

D_{zw}

-0.0108
-0.01041

C_y

0	0	0	5.066579	-116.446	0	0	0
0	0	0	0	1	0	0	0
12.96989	10.32532	-0.56926	0	0	0	0	0

D_{yu}

0
0
0

D_{yw}

0
0
-0.29656

The eigenvalues of the FCS linear plant are

$$\lambda = \begin{bmatrix} -219.63 & -89.485 & -46.177 & -22.404 & -18.265 & -2.9161 & -1.6474 & -1.4038 \end{bmatrix} \tag{A.1}$$

The required compressor flow rate that satisfies the desired oxygen excess ratio can be calculated from

Table A.2. Linearization results including static feedforward

A

-6.30908	0	-10.9544	0	83.74458	0	0	24.05866
0	-161.083	0	0	51.52923	0	-18.0261	0
-18.7858	0	-46.3136	0	275.6592	0	0	158.3741
0	0	0	-17.3506	193.9373	0	0	0
1.299576	0	2.969317	0.3977	-38.7024	0.105748	0	0
16.64244	0	38.02522	5.066579	-479.384	0	0	0
0	-450.386	0	0	142.2084	0	-80.9472	0
2.02257	0	4.621237	0	0	0	0	-51.2108

B_u

0
0
0
3.942897
0
0
0
0

B_w

-0.03159
-0.00398
0
2.681436
0
0
-0.05242
0

C_z

-2.48373	-1.9773	0.109013	-0.21897	0	0	0	0
-0.63477	0	-1.45035	0	13.84308	0	0	0

D_{zu}

0.168979
0

D_{zw}

0.104116
-0.01041

C_y

0	0	0	5.066579	-116.446	0	0	0
0	0	0	0	1	0	0	0
12.96989	10.32532	-0.56926	0	0	0	0	0

D_{yu}

0
0
0

D_{yw}

0
0
-0.29656

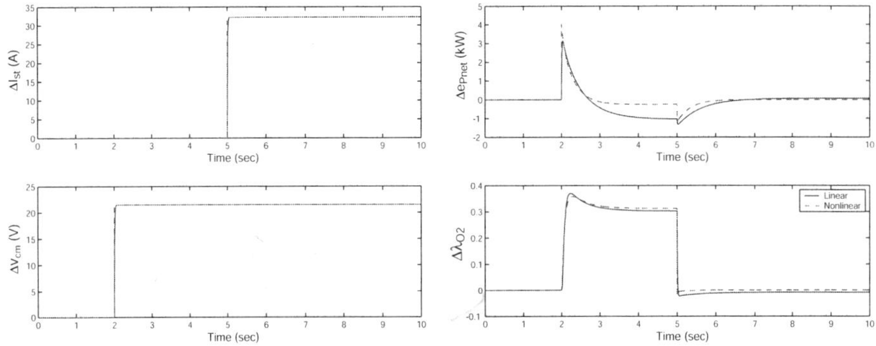

Fig. A.1. Comparison of FCS responses between nonlinear and linear models

$$W_{cp} = (1 + \omega_{atm}) W_{air} \tag{A.2}$$

$$= \left(1 + \frac{M_v}{M_a} \frac{p_{sat}(T_{atm})}{p_{atm} - p_{sat}(T_{atm})}\right) \frac{1}{x_{O_2}} \lambda_{O_2} M_{O_2} \frac{nI}{4F} \tag{A.3}$$

where ω_{atm} is the humidity ratio of the atmospheric air, W_{air} is the required mass flow rate of dry air, M_a is the dry air molar mass, and x_{O_2} is the oxygen mass fraction in dry air. If the mole fraction of oxygen in the dry air is 0.21, the values of M_a and x_{O_2} are 28.84×10^{-3} kg/mol and 0.23301, respectively.

The observer gain here is used in (5.19) for the FCS system observer.

$$L = \begin{bmatrix} 17.667 & -0.31359 & 2785.1 \\ -9.1023 & 5.9184 & 20.78 \\ 23.741 & 99.245 & -247.9 \\ 8159.8 & 446.15 & 1496 \\ -20.122 & 19.43 & 54.645 \\ -155.13 & 265.91 & 1057 \\ -72.553 & 5.7114 & -31.406 \\ 7.2343 & 6.6423 & 65.186 \end{bmatrix} \qquad (A.4)$$

A.2 FPS Control Design

Table A.3 shows the matrices of the linearized FPS model with static feedforward included and Figure A.2 shows the responses from both the linear and nonlinear models.

Table A.3. FPS linear model system matrices

A

-0.074	0	0	0	0	0	-3.53	1.0748	0	1E-06
0	-1.468	-25.3	0	0	0	0	0	2.5582	13.911
0	0	-156	0	0	0	0	0	0	33.586
0	0	0	-124.5	212.63	0	112.69	112.69	0	0
0	0	0	0	-3.333	0	0	0	0	0
0	0	0	0	0	-32.43	32.304	32.304	0	0
0	0	0	0	0	331.8	-344	-341	0	9.9042
0	0	0	221.97	0	0	-253.2	-254.9	0	32.526
0	0	2.0354	0	0	0	1.8309	1.214	-0.358	-3.304
0.0188	0	8.1642	0	0	0	5.6043	5.3994	0	-13.61

Bu

0	0
0	0
0	0
0	0
0.12	0
0	0.1834
0	0
0	0
0	0
0	0

Bw

0
-0.328
-0.024
0
0.0265
0.0504
0
0
0
0

Cz

1	0	0	0	0	0	0	0	0	0
0	0.994	-0.088	0	0	0	0	0	0	0

Dzu

0	0
0	0

Dzw

0
0

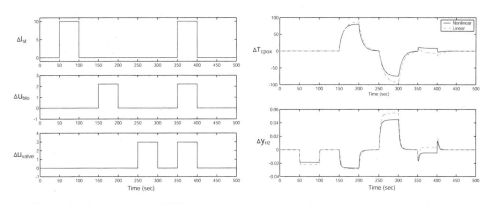

Fig. A.2. Comparison of FPS responses between nonlinear and linear models

The closed loop eigenvalues of the FPS system in Section 7.5.1 are

$$\lambda(A_a - B_a K) = \begin{bmatrix} -660.66 \\ -157.92 \\ -89.097 \\ -12.174 \\ -5.8954 \\ -3.0122 + 3.8238i \\ -3.0122 - 3.8238i \\ -1.3514 + 0.6979i \\ -1.3514 - 0.6979i \\ -1.3578 \\ -0.27348 \\ -0.35855 \end{bmatrix} \qquad (A.5)$$

Table A.4 shows the FPS system matrices after the state is transformed to the modal canonical form.

Table A.4. FPS linear model system matrices in modal canonical form

A_1

										B_{1u}		B_{1w}
-660.7	0	0	0	0	0	0	0	0	0	-0.011	0.0646	0.0154
0	-89.1	0	0	0	0	0	0	0	0	-0.661	-0.386	-0.251
0	0	-157.9	0	0	0	0	0	0	0	-0.001	0.0004	-0.024
0	0	0	-12.17	0	0	0	0	0	0	-0.155	0.4411	0.0937
0	0	0	0	-3.333	0	0	0	0	0	368.64	0	81.314
0	0	0	0	0	-2.782	0.4612	0	0	0	345.79	-4.027	75.166
0	0	0	0	0	-0.461	-2.782	0	0	0	-378.3	-15.9	-87.76
0	0	0	0	0	0	0	-1.468	0	0	-3.555	-1.855	-1.618
0	0	0	0	0	0	0	0	-0.358	0	-0.568	0.9745	0.1421
0	0	0	0	0	0	0	0	0	-0.086	3.2203	-2.916	-0.09

C_{1z}

										D_{1zu}		D_{1zw}
-0.003	0.0354	-3E-05	-0.248	0.6991	-0.755	0	0	0	0.9999	0	0	0
-2E-04	0.0008	0.0764	0.1414	0.0895	-0.113	-0.027	0.994	0.9119	-0.006	0	0	0

The observer gain for the FPS estimator with perfect measurements in Section 7.5.2 is

$$L = \begin{bmatrix} 469.67 & 138.59 \\ 5.6245 & 818.34 \\ -93.622 & -12.759 \\ -742.29 & -99.077 \\ -30.245 & -4.7928 \\ -795.89 & -104.2 \\ -2149.2 & -392.69 \\ 1400.7 & 294.09 \\ 1559.8 & 3547.1 \\ -430.89 & -58.728 \end{bmatrix} \qquad (A.6)$$

The C and D matrices for the FPS system with realistic measurements are given in (A.7).

$$C_y = \begin{bmatrix} 0\;0\;0\;0\;0\;0 & 0 & 0 & 0\;0\;1\;0 \\ 0\;0\;0\;0\;0\;0 & 0 & 0 & 0\;0\;0\;1 \\ 0\;0\;0\;0\;0\;0 & -4.2747 & 7.17 & 0\;0\;0\;0 \\ 0\;0\;0\;0\;0\;0 & 0 & -0.1350 & 0\;0\;0 \end{bmatrix} \tag{A.7}$$

$$Dyu = \begin{bmatrix} 0 & 0 \\ 0 & 0 \\ 0 & 0 \\ 0 & 0.202 \end{bmatrix} \quad Dyw = \begin{bmatrix} 0 \\ 0 \\ 0 \\ 0.0555 \end{bmatrix}$$

References

[1] Adams, J., Yang, W.-C., Oglesby, K., and Osborne, K. (2000). The development of Ford's P2000 fuel cell vehicle. *SAE Paper 2000-01-1061.*

[2] Ahmed, S. and Krumpelt, M. (2001). Hydrogen from hydrocarbon fuels for fuel cells. *International Journal of Hydrogen Energy*, **26**, 291–301.

[3] Akella, S., Sivashankar, N., and Gopalswamy, S. (2001). Model-based systems analysis of a hybrid fuel cell vehicle configuration. *Proceedings of 2001 American Control Conference*, **3**, 1777–1782.

[4] Amphlett, J., Baumert, R., Mann, R., Peppley, B., Roberge, P., and Rodrigues, A. (1994). Parametric modelling of the performance of a 5-kW proton-exchange membrane fuel cell stack. *Journal of Power Sources*, **49**, 349–356.

[5] Amphlett, J., Baumert, R., Mann, R., Peppley, B., and Roberge, P. (1995). Performance modeling of the Ballard Mark IV solid polymer electrolyte fuel cell. *Journal of Electrochemical Society*, **142**(1), 9–15.

[6] Amphlett, J., Mann, R., Peppley, B., Roberge, P., and Rodrigues, A. (1996). A model predicting transient responses of proton exchange membrane fuel cells. *Journal of Power Sources*, **61**, 183–188.

[7] Anderson, B. and Moore, J. (1989). *Optimal Control: Linear Quadratic Methods*. Prentice-Hall, Englewood Cliffs, NJ.

[8] Appleby, A. and Foulkes, F. (1989). *Fuel Cell Handbook*. Van Nostrand Reinhold, New York.

[9] Arcak, M., Gorgun, H., Pedersen, L., and Varigonda, S. (2003). An adaptive observer design for fuel cell hydrogen estimation. *Proceedings of the 2003 American Control Conference*, pages 2037–2042.

[10] Atwood, P., Gurski, S., Nelson, D., Wipke, K., and Markel, T. (2001). Degree of hybridization ADVISOR modeling of a fuel cell hybrid electric sport utility vehicle. *Proceedings of 2001 Joint ADVISOR/PSAT Vehicle Systems Modeling User Conference*, pages 147–155.

[11] Badrinarayanan, P., Ramaswamy, S., Eggert, A., and Moore, R. (2001). Fuel cell stack water and thermal management: Impact of variable system power operation. *SAE Paper 2001-01-0537.*

[12] Barbir, F., Balasubramanian, B., and Neutzler, J. (1999). Trade-off design analysis of operating pressure and temperature in PEM fuel cell systems. *Proceedings of the ASME Advanced Energy Systems Division*, **39**, 305–315.

[13] Barbir, F., Fuchs, M., Husar, A., and Neutzler, J. (2000). Design and operational characteristics of automotive PEM fuel cell stacks. *SAE Paper 2000-01-0011*.

[14] Baschuk, J. and Li, X. (2000). Modelling of polymer electrolyte membrane fuel cells with variable degrees of water flooding. *Journal of Power Sources*, **86**, 186–191.

[15] Bernardi, D. (1990). Water-balance calculations for solid-polymer-electrolyte fuel cells. *Journal of Electrochemical Society*, **137**(11), 3344–3350.

[16] Bernardi, D. and Verbrugge, M. (1992). A mathematical model of the solid-polymer-electrolyte fuel cell. *Journal of the Electrochemical Society*, **139**(9), 2477–2491.

[17] Bevers, D., Wöhr, M., Yasuda, K., and Oguro, K. (1997). Simulation of a polymer electrolyte fuel cell electrode. *Journal of Applied Electrochemistry*, **27**(11), 1254–1264.

[18] Birch, S. (2001). Ford's focus on the fuel cell. *Automotive Engineering International*, pages 25–28.

[19] Boettner, D., Paganelli, G., Guezennec, Y., Rizzoni, G., and Moran, M. (2001a). Component power sizing and limits of operation for proton exchange membrane (PEM) fuel cell/battery hybrid automotive applications. *Proceedings of 2001 ASME International Mechanical Engineering Congress and Exposition*.

[20] Boettner, D., Paganelli, G., Guezennec, Y., Rizzoni, G., and Moran, M. (2001b). Proton exchange membrane (PEM) fuel cell system model for automotive vehicle simulation and control. *Proceedings of 2001 ASME International Mechanical Engineering Congress and Exposition*.

[21] Boyce, M. (1982). *Gas Turbine Engineering Handbook*. Gulf, Houston, TX.

[22] Bristol, E. (1966). On a new measure of interactions for multivariable process control. *IEEE Transactions on Automatic Control*, **AC-11**, 133–134.

[23] Brown, L. (2001). A comparative study of fuels for on-board hydrogen production for fuel-cell-powered automobiles. *International Journal of Hydrogen Energy*, **26**, 381–397.

[24] Büchi, F. and Srinivasan, S. (1997). Operating proton exchange membrane fuel cells without external humidification of the reactant gases. *Journal of Electrochemical Society*, **144**(8), 2767–2772.

[25] Chan, S. and Wang, H. (2000). Thermodynamic analysis of natural-gas fuel processing for fuel cell applications. *International Journal of Hydrogen Energy*, **25**, 441–449.

[26] Chen, C.-T. (1998). *Linear System Theory and Design*. Oxford University Press, New York.

[27] Chu, D. and Jiang, R. (1999a). Comparative studies of polymer electrolyte membrane fuel cell stack and single cell. *Journal of Power Sources*, **80**, 226–234.

[28] Chu, D. and Jiang, R. (1999b). Performance of polymer electrolyte membrane fuel cell (PEMFC) stacks, part I. Evaluation and simulation of an air-breathing PEMFC stack. *Journal of Power Sources*, **83**, 128–133.

[29] Cunningham, J., Hoffman, M., Moore, R., and Friedman, D. (1999). Requirements for a flexible and realistic air supply model for incorporation into a fuel cell vehicle (FCV) system simulation. *SAE Paper 1999-01-2912*.

[30] Dannenberg, K., Ekdunge, P., and Lindbergh, G. (2000). Mathematical model of the PEMFC. *Journal of Applied Electrochemistry*, **30**, 1377–1387.

[31] Davis, S. (2000). *Transportation Energy Data Book*. U.S. Department of Energy, 20 edition. ORNL-6959.

[32] de Smet, C., de Croon, M., Berger, R., Marin, G., and Schouten, J. (2001). Design of adiabatic fixed-bed reactors for the partial oxidation of methane to synthesis gas. Application to production of methanol and hydrogen-for-fuel-cells. *Chemical Engineering Science*, **56**, 4849–4861.

[33] Devasia, S. (2002). Should model-based inverse inputs be used as feedforward under plant uncertainty? *IEEE Transactions on Automatic Control*, **41**(11), 1865–1871.

[34] Devasia, S., Chen, D., and Paden, B. (1996). Nonlinear inversion-based output tracking. *IEEE Transactions on Automatic Control*, **41**(7), 930–942.

[35] Di Benedetto, M. and Lucibello, P. (1993). Inversion of nonlinear time-varying systems. *IEEE Transactions on Automatic Control*, **38**(8), 1259–1264.

[36] Dicks, A. (1996). Hydrogen generation from natural gas for the fuel cell systems of tomorrow. *Journal of Power Sources*, **61**, 113–124.

[37] Doss, E., Kumar, R., Ahluwalia, R., and Krumpelt, M. (2001). Fuel processors for automotive fuel cell systems: a parametric analysis. *Journal of Power Sources*, **102**, 1–15.

[38] Doyle, J. and Stein, G. (1979). Robustness with observers. *Transactions on Automatic Control*, **AC-24**(4), 607–611.

[39] Dutta, S., Shimpalee, S., and Van Zee, J. (2001). Numerical prediction of mass-exchange between cathode and anode channels in a PEM fuel cell. *International Journal of Heat and Mass Transfer*, **44**, 2029–2042.

[40] Eborn, J., Pedersen, L., Haugstetter, C., and Ghosh, S. (2003). System level dynamic modeling of fuel cell power plants. *Proceedings of the 2003 American Control Conference*, pages 2024–2029.

[41] Francis, B. and Wonham, W. (1976). The internal model principle of control theory. *Automatica*, **12**(5), 457–465.

[42] Franklin, G., Powell, J., and Emani-Naeini, A. (1994). *Feedback Control of Dynamic Systems*. Addison-Wesley, New York.

[43] Freudenberg, J. (2002). A first graduate course in feedback control. University of Michigan, EECS 565 Coursepack.

[44] Freudenberg, J. and Middleton, R. (1999). Properties of single input, two output feedback systems. *International Journal of Control*, **72**(16), 1446–1465.

[45] Friedman, D., Egghert, A., Badrinarayanan, P., and Cunningham, J. (2001). Balancing stack, air supply and water/thermal management demands for an indirect methanol PEM fuel cell system. *SAE Paper 2001-01-0535*.

[46] Fronk, M., Wetter, D., Masten, D., and Bosco, A. (2000). PEM fuel cell system solutions for transportation. *SAE Paper 2000-01-0373*.

[47] Fuchs, M., Barbir, F., Husar, A., Neutzler, J., Nelson, D., Ogburn, M., and Bryan, P. (2000). Performance of automotive fuel cell stack. *SAE Paper 2000-01-1529*.

[48] Fuller, T. and Newman, J. (1993). Water and thermal management in solid-polymer-electrolyte fuel cells. *Journal of Electrochemical Society*, **140**(5), 1218–1225.

[49] Gardner, T., Berry, D., Lyons, K., Beer, S., and Freed, A. (2002). Fuel processor integrated H_2S catalytic partial oxidation technology for sulfur removal in fuel cell power plants. *Fuel*, **81**, 2157–2166.

[50] Georgiou, T. and Smith, M. (1990). Optimal robustness in the gap metric. *IEEE Transactions on Automatic Control*, **35**(6), 673–686.

[51] Geyer, H., Ahluwalia, R., and Kumar, R. (1996). Dynamic response of steam-reformed, methanol-fueled, polymer electrolyte fuel cell systems. *Proceedings of the Intersociety Energy Conversion Engineering Conference*, **2**, 1101–1106.

[52] Glass, R., Milliken, J., Howden, K., and Sullivan, R. (2000). *Sensor Needs and Requirements for Proton Exchange Membrane Fuel Cell Systems and Direct-Injection Engines*. Department of Energy. Published by Lawrence Livermore National Laboratory.

[53] Gravdahl, J. and Egeland, O. (1999). *Compressor Surge and Rotating Stall*. Springer, London.

[54] Grove, W. (1839). A small voltaic battery of great energy. *Philosophical Magazine*, **15**, 287–293.

[55] Gurau, V., Liu, H., and Kakac, S. (1998). Mathematical model for proton exchange membrane fuel cells. *Proceedings of the 1998 ASME Advanced Energy Systems Division*, pages 205–214.

[56] Guzzella, L. (1999). Control oriented modelling of fuel-cell based vehicles. *Presentation in NSF Workshop on the Integration of Modeling and Control for Automotive Systems*.

[57] Hauer, K.-H., Friedmann, D., Moore, R., Ramaswamy, S., Eggert, A., and Badrinarayana, P. (2000). Dynamic response of an indirect-methanol fuel cell vehicle. *SAE Paper 2000-01-0370*.

[58] Heywood, J. (1988). *Internal Combustion Engine Fundamentals*. McGraw-Hill, New York.

[59] Hunter, G., Backford, R., Jansa, E., Makel, D., Liu, C., Wu, Q., and Powers, W. (1994). Microfabricated hydrogen sensor technology for aerospace and commercial applications. *Presented at NASA/SPIE Symposium, San Diego*.

[60] Jiang, R. and Chu, D. (2001a). Stack design and performance of polymer electrolyte membrane fuel cells. *Journal of Power Sources*, **93**, 25–31.

[61] Jiang, R. and Chu, D. (2001b). Voltage-time behavior of a polymer electrolyte membrane fuel cell stack at constant current discharge. *Journal of Power Sources*, **92**, 193–198.

[62] Jost, K. (2000). Fuel-cell concepts and technology. *Automotive Engineering International*.

[63] Kailath, T. (1980). *Linear Systems*. Prentice-Hall, Englewood Cliffs, NJ.

[64] Kalhammer, F., Prokopius, P., Roan, V., and Voecks, G. (1998). *Status and prospects of fuel cells as automobile engines*. State of California Air Resources Board.

[65] Kim, J., Lee, S.-M., and Srinivasan, S. (1995). Modeling of proton exchange membrane fuel cell performance with an empirical equation. *Journal of the Electrochemical Society*, **142**(8), 2670–2674.

[66] Kim, Y.-H. and Kim, S.-S. (1999). An electrical modeling and fuzzy logic control of a fuel cell generation system. *IEEE Transactions of Energy Conversion*, **14**(2), 239–244.

[67] Kordesch, K. and Simader, G. (1996). *Fuel Cells and Their Applications*. VCH, Weinheim, Germany.

[68] Larentis, A., de Resende, N., Salim, V., and Pinto, J. (2001). Modeling and optimization of the combined carbon dioxide reforming and partial oxidation of natural gas. *Applied Catalysis*, **215**, 211–224.

[69] Larminie, J. and Dicks, A. (2000). *Fuel Cell Systems Explained*. Wiley, West Sussex, England.

[70] Laurencelle, F., Chahine, R., Hamelin, J., Agbossou, K., Fournier, M., Bose, T., and Laperriere, A. (2001). Characterization of a Ballard MK5-E proton exchange membrane fuel cell stack. *Fuel Cells Journal*, **1**(1), 66–71.

[71] Ledjeff-Hey, K., Roses, J., and Wolters, R. (2000). CO_2-scrubbing and methanation as purification system for PEFC. *Journal of Power Sources*, **86**, 556–561.

[72] Lee, J. and Lalk, T. (1998). Modeling fuel cell stack systems. *Journal of Power Sources*, **73**, 229–241.

[73] Lee, J., Lalk, T., and Appleby, A. (1998). Modeling electrochemical performance in large scale proton exchange membrane fuel cell stacks. *Journal of Power Sources*, **70**, 258–268.

[74] Lesster, L. (2000). Fuel cell power electronics: Managing a variable-voltage DC source in a fixed-voltage AC world. *Fuel Cells Bulletin*, **2**(25), 5–9.

[75] Lorenz, H., Noreikat, K.-E., Klaiber, T., Fleck, W., Sonntag, J., Hornburg, G., and Gaulhofer, A. (1997). Method and device for vehicle fuel cell dynamic power control. *United States Patents 5,646,852*.

[76] Maggio, G., Recupero, V., and Pino, L. (2001). Modeling polymer electrolyte fuel cells: An innovative approach. *Journal of Power Sources*, **101**, 275–286.

[77] Mann, R., Amphlett, J., Hooper, M., Jensen, H., Peppley, B., and Roberge, P. (2000). Development and application of a generalized steady-state electrochemical model for a PEM fuel cell. *Journal of Power Sources*, **86**, 173–180.

[78] Marr, C. and Li, X. (1998). Performance modelling of a proton exchange membrane fuel cell. *Proceedings of Energy Sources Technology Conference and Exhibition*, pages 1–9.

[79] Megede, D. (2002). Fuel processors for fuel cell vehicles. *Journal of Power Sources*, **106**, 35–41.

[80] Moraal, P. and Kolmanovsky, I. (1999). Turbocharger modeling for automotive control applications. *SAE Paper 1999-01-0908*.

[81] Morari, M. and Zafiriou, E. (1997). *Robust Process Control*. Prentice Hall, Englewood Cliffs, NJ.

[82] Mufford, W. and Strasky, D. (1999). Power control system for a fuel cell powered vehicle. *United States Patents 5,991,670*.

[83] Nguyen, T. and White, R. (1993). A water and heat management model for proton-exchange-membrane fuel cells. *Journal of Electrochemical Society*, **140**(8), 2178–2186.

[84] Ogburn, M., Nelson, D., Wipke, K., and Markel, T. (2000a). Modeling and validation of a fuel cell hybrid vehicle. *SAE Paper 2000-01-1566*.

[85] Ogburn, M., Nelson, D., Luttrell, W., King, B., Postle, S., and Fahrenkrog, R. (2000b). Systems integration and performance issues in a fuel cell hybrid electric vehicle. *SAE Paper 2000-01-0376*.

[86] Ogunnaike, B. and Ray, W. (1994). *Process Dynamics, Modeling, and Control*. Oxford University Press, New York.

[87] Okada, T., Xie, G., and Meeg, M. (1998). Simulation for water management in membranes for polymer electrolyte fuel cells. *Electrochimica Acta*, **43**.

[88] Padulles, J., Ault, G., Smith, C., and McDonald, J. (1999). Fuel cell plant dynamic modelling for power systems simulation. *Proceedings of 34th Universities Power Engineering Conference*, **34**(1), 21–25.

[89] Paganin, V., Oliveira, C., Ticianelli, E., Springer, T., and Gonzalez, E. (1998). Modelistic interpretation of the impedance response of a polymer electrolyte fuel cell. *Electrochimica Acta*, **43**(24), 3761–3766.

[90] Panik, F. (1998). Fuel cells for vehicle applications in cars – bringing the future closer. *Journal of Power Sources*, **71**, 36–38.

[91] Pino, L., Recupero, V., Beninati, S., Shukla, A., Hegde, M., and Bera, P. (2002). Catalytic partial-oxidation of methane on a ceria-supported platinum catalyst for application in fuel cell electric vehicles. *Applied Catalysis A: General*, **225**, 63–75.

[92] Pischinger, S., Schönfelder, C., Bornscheuer, W., Kindl, H., and Wiartalla, A. (2001). Integrated air supply and humidification concepts for fuel cell systems. *SAE Paper 2001-01-0233*.

[93] Pukrushpan, J., Stefanopoulou, A., and Peng, H. (2002). Modeling and control issues of PEM fuel cell stack system. *Proceedings of the 2002 American Control Conference*, pages 3117–3122.

[94] Rajashekara, K. (2000). Propulsion system strategies for fuel cell vehicles. *SAE Paper 2000-01-0369*.

[95] Recupero, V., Pino, L., Leonardo, R., Lagana, M., and Maggio, G. (1998). Hydrogen generator, via catalytic partial oxidation of methane for fuel cells. *Journal of Power Sources*, **71**, 208–214.

[96] Ro, K. and Rahman, S. (2003). Control of grid-connected fuel cell plants for enhancement of power system stability. *Renewable Energy*, **28**, 397–407.

[97] Rodatz, P. (2003). *Dynamics of the polymer electrolyte fuel cell: Experiments and model-based analysis*. Ph.D. thesis, Swiss Federal Institute of Technology, Zurich.

[98] Sadler, M., Stapleton, A., Heath, R., and Jackson, N. (2001). Application of modeling techniques to the design and development of fuel cell vehicle systems. *SAE Paper 2001-01-0542*.

[99] Scheiber, S. (2003). Too hot to handle. *Control Engineering Magazine*, **50**(6).

[100] Singh, D., Lu, D., and Djilali, N. (1999). A two-dimensional analysis of mass transport in proton exchange membrane fuel cells. *International Journal of Engineering Science*, **37**, 431–452.

[101] Skogestad, S. and Postlethwaite, I. (1996). *Multivariable Feedback Control: Analysis and Design*. Wiley, West Sussex, England.

[102] Song, R.-H., Kim, C.-S., and Shin, D. (2000). Effects of flow rate and starvation of reactant gases on the performance of phosphoric acid fuel cells. *Journal of Power Sources*, **86**, 289–293.

[103] Sonntag, R., Borgnakke, C., and Van Wylen, G. (1998). *Fundamentals of Thermodynamics*. Wiley, New York, fifth edition.

[104] Springer, T., Zawodzinski, T., and Gottesfeld, S. (1991). Polymer electrolyte fuel cell model. *Journal of Electrochemical Society*, **138**(8), 2334–2342.

[105] Springer, T., Rockward, R., Zawodzinski, T., and Gottesfeld, S. (2001). Model for polymer electrolyte fuel cell operation on reformate feed. *Journal of The Electrochemical Society*, **148**, A11–A23.

[106] Sridhar, P., Perumal, R., Rajalakshmi, N., Raja, M., and Dhathathreyan, K. (2001). Humidification studies on polymer electrolyte membrane fuel cell. *Journal of Power Sources*, **101**, 72–78.

[107] Stobart, R. (1999). Fuel cell power for passenger cars – what barriers remain? *SAE Paper 1999-01-0321*.

[108] The Argus Group (2001). Hydrogen sensor for automotive fuel cells from the Argus Group. *http://www.fuelcellsensor.com/*.

[109] The MathWorks, Inc. (2001). *μ-Analysis and Synthesis Toolbox Users Guide*. Natick, MA.

[110] Thirumalai, D. and White, R. (1997). Mathematical modeling of proton-exchange-membrane fuel-cell stacks. *Journal of Electrochemical Society*, **144**(5).

[111] Thomas, C., James, B., Lomax, Jr., F., and Kuhn, Jr., I. (2000). Fuel options for the fuel cell vehicle: Hydrogen, methanol or gasoline? *International Journal of Hydrogen Energy*, **25**, 551–567.

[112] Thomas, S. and Zalbowitz, M. (2000). *Fuel Cells Green Power*. Los Alamos National Laboratory, Los Alamos, NM.

[113] Thorstensen, B. (2001). A parametric study of fuel cell system efficiency under full and part load operation. *Journal of Power Sources*, **92**, 9–16.

[114] Tiller, M. (2001). *Introduction to Physical Modeling with Modelica*. Kluwer Academic, Boston.

[115] Turner, W., Parten, M., Vines, D., Jones, J., and Maxwell, T. (1999). Modeling a PEM fuel cell for use in a hybrid electric vehicle. *Proceedings of the 1999 IEEE 49th Vehicular Technology Conference*, **2**, 1385–1388.

[116] U.S. Department of Energy, Office of Fossil Energy, and National Energy Technology Laboratory (2002). *Fuel Cell Handbook*. EG&G Technical Services, Inc, Science Application International Corporation.

[117] U.S. Environmental Protection Agency (2000). *Latest Findings on National Air Quality: 1999 Status and Trends*. EPA-454/F-00-002.

[118] Wagner, N., Schnurnberger, W., Muller, B., and Lang, M. (1998). Electrochemical impedance spectra of solid-oxide fuel cells and polymer membrane fuel cells. *Electrochimica Acta*, **43**(24), 3785–3793.

[119] Wöhr, M., Bolwin, K., Schnurnberger, W., Fischer, M., Neubrand, W., and Eigenberger, G. (1998). Dynamic modelling and simulation of a polymer membrane fuel cell including mass transport limitation. *International Journal for Hydrogen Energy*, **23**(3), 213–218.

[120] Yang, W.-C., Bates, B., Fletcher, N., and Pow, R. (1998). Control challenges and methodologies in fuel cell vehicle development. *SAE Paper 98C054*.

[121] Yanga, T.-H., Yoona, Y.-G., Kima, C.-S., Kwakb, S.-H., and Yoon, K.-H. (2002). A novel preparation method for a self-humidifying polymer electrolyte membrane. *Journal of Power Sources*, **102**, 328–332.

[122] Zhu, J., Zhang, D., and King, K. (2001). Reforming of CH_4 by partial oxidation: Thermodynamic and kinetic analyses. *Fuel*, **80**, 899–905.

Index